924

Elementary
Problems and Answers
in Solar System Astronomy

This diagram shows the path (right ascension α and declination δ) of Mars on the star field during 1986 according to *The Astronomical Almanac*. Mars was in opposition on 7/10 and stationary on 6/10 and 8/12. The loop epitomizes the classical problem of understanding planetary motion (cf. problem 42).

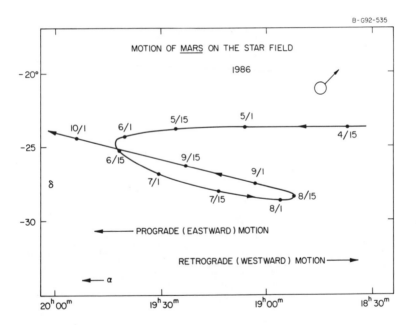

924

Elementary
Problems and Answers
in Solar System Astronomy

James A. Van Allen

Iowa City
University of Iowa Press

University of Iowa Press, Iowa City 52242
Printed in the United States of America

This book was printed from camera-ready copy
prepared under the direction of the author.

Jacket/cover: This portion of the equatorial sky
was photographed with a fixed camera by a team
of students in the author's 1979–80 general
astronomy class—John Finnegan, Scott Gallup,
Larry Granroth, and Dave Peters. The open
dome of the University of Iowa's Hills
Observatory is in the foreground. Two problems
for the reader: (a) Identify the bright stars in the
constellation Orion (at right) and determine the
duration of the time exposure. (b) Identify the
bright star at the upper left.

Library of Congress Cataloging-in-
Publication Data

Van Allen, James Alfred, 1916–
924 elementary problems and answers in
solar system astronomy / by James A.
Van Allen.
p.   cm.
Includes bibliographical references.
ISBN 0-87745-433-7, ISBN
0-87745-434-5 (pbk.)
1. Solar system—Miscellanea.   I.
Title.   II. Title: Nine hundred twenty-
four elementary problems and answers in
solar system astronomy.
QB502.V36   1993   93-24605
523.2—dc20   CIP

Printed on acid-free paper

97 96 95 94 93 C 5 4 3 2 1
97 96 95 94 93 P 5 4 3 2 1

☆　☆　☆

Contents

Preface

This book contains 924 problems, gleaned from quizzes, examinations, observational exercises, classroom discussions, and special assignments that I prepared during my teaching of an introductory course in Solar System Astronomy for seventeen years at the University of Iowa. The problems are loosely organized into seven chapters. They range in difficulty from ones requiring only simple factual knowledge to ones requiring significant understanding and analysis. Some are essentially repetitive but are recast in different forms. An answer is given to each problem with the exception of a few that require personal observation.

I hope that this collection of problems will challenge beginning students of astronomy as well as self-taught amateurs and that it will be useful to hard-pressed instructors of courses comparable to mine. For instructors who have a justifiable aversion to multiple choice questions, I may note that most of my problems of this type can be readily converted to ones requiring short answers or reasoned discussion.

My principal profession is that of a physicist, having had only one formal course in astronomy, as an undergraduate student in 1934 at Iowa Wesleyan College. I was freshly inspired to learn and teach Solar System Astronomy by my progressive involvement in interplanetary and planetary research beginning in 1958 *and* by the successive editions of George O. Abell's excellent textbook *Exploration of the Universe*. My more recent participation in the extended space missions of Pioneer 10 and Pioneer 11 under the auspices of the Ames Research Center of the National Aeronautics and Space Administration has sustained my working dedication to the subject.

I am especially indebted to Evelyn D. Robison and Alice M. Shank for working with me through an uncounted number of revisions and corrections and for preparation of the camera-ready copy of the completed manuscript. The figures were drawn by the skilled hand of John R. Birkbeck.

James A. Van Allen
Regent Distinguished Professor
University of Iowa
March 1993

☆ ☆ ☆

CHAPTER 1

The Sun and the Nature
of the Solar System

1. List characteristics of the solar system that are major clues in devising an hypothesis of its origin and evolution.

2. The theory of origin of the solar system most widely accepted by professional astronomers at present is principally due to (a) Velikovsky; (b) Menzel; (c) Oort; (d) Kuiper.

3. In the nebular theory of the origin of the solar system, it is visualized that all of the planets (a) were developed by accretion from a primordial cloud of gas and dust; (b) are fragments of the Sun that resulted from disruptive tidal forces of a passing star; (c) have the same average chemical composition; (d) are captured bodies that were moving previously through interstellar space.

4. The general theory of stellar evolution indicates that (a) the Sun will change to a pulsating star within about one million years from now; (b) the Sun will have about its present size and brightness for at least another billion years; (c) the Sun's brightness will decrease at the rate of about one percent per year during the next 100 years; (d) the Sun's brightness will increase by about one percent per year during the next 100 years.

5. Current theoretical models of the interior of the Sun have *all but one* of the following features: (a) The temperature at the center is enormously greater than at the surface of the photosphere. (b) Elements such as iron, nickel, chromium, and molybdenum are concentrated near the surface. (c) Most of the atoms in the Sun are partially or completely ionized. (d) The density at the center is about 160 g cm^{-3}.

6. The Sun (a) is the largest object in the solar system; (b) is an excellent example of a pulsar; (c) is properly classified as a very large planet; (d) has an estimated lifetime of about three million years.

7. It is generally considered by astronomers (a) that few, if any, stars other than the Sun have planets; (b) that Kepler's laws are not applicable outside of the solar system; (c) that there probably are planetary systems associated with many millions of other stars.

8. The principal source of the energy emitted by the Sun is (a) decay of radioactive substances; (b) chemical combustion; (c) gravitational contraction; (d) nuclear fusion of hydrogen to form helium.

9. At 1 AU (150,000,000 km) sunlight provides a power flow of 1.4 kilowatts per square meter. Hence the *total* power flow from the Sun in all directions is _____ kilowatts.

10. What is the basis for the claim that hydrogen is the principal constituent of the Sun?

11. Most of the light from the Sun comes from (a) the corona; (b) flares; (c) the photosphere; (d) the chromosphere.

12. The temperature of the Sun's photosphere is about (a) 1,000,000; (b) 15,000,000; (c) 20,000; (d) 5,800 degrees Kelvin.

13. The solar spectrum has its greatest intensity (a) at radio frequencies; (b) in the infrared portion of the spectrum; (c) in the blue-green portion of the spectrum; (d) in the ultraviolet portion of the spectrum.

14. Fraunhofer lines are (a) thin, bright filaments seen in photographs of the Sun in the light of atomic hydrogen; (b) emission lines in the spectrum of the whole disc of the Sun; (c) emission lines in the coronal spectrum as observed during total eclipse of the Sun; (d) absorption lines of many different elements in the spectrum of the whole disc of the Sun.

15. The presence of specific chemical elements in the solar atmosphere is revealed by what features in the spectrum of the Sun?

16. The atomic composition of the outer layers of the Sun has been learned principally from study of (a) absorption lines in the spectrum of photospheric light; (b) the application of Wien's law to the solar spectrum; (c) the Doppler shift of the Hα emission line in solar flares; (d) the average density of the Sun, taken as a whole.

17. Under the assumptions that the Sun is spherically symmetrical and that its density declines linearly with radial distance from some specific value at the center to zero at the top of the photosphere, find the radius of the spherical shell that encloses 50 percent of the mass. Call the radius of the top of the photosphere 1.00.

18. What is the approximate distance in AU from the Sun to the nearest star, Proxima Centauri?

19. Define the effective gravitational radius of the solar system and estimate its magnitude.

20. The angular radius of the Sun as viewed from the Earth is 16 minutes of arc. Hence, the distance from the Earth to the Sun is _____ times as great as the radius of the Sun.

21. The Sun has a radius one hundred ten times greater than that of the Earth and a mean density one-fourth as great. With these approximate data, the Sun's mass is found to be (a) 1,330,000; (b) 330,000; (c) 25,000; (d) 3,000 times as great as that of the Earth.

22. The chromosphere of the Sun is (a) a spherical shell of gas extending to many millions of km; (b) a spherical shell of gas about 3,000 km thick; (c) that part of the Sun that emits most of the energy that is observable at the Earth; (d) the most extended part of a total solar eclipse photograph.

23. The radioactive half life for the decay of Rb^{87} to Sr^{87} is 5.0×10^{10} years. During the lifetime of the solar system about what percentage of the original rubidium has decayed?

24. The temperature of the gas in the solar corona is (a) about 1 million; (b) over 10 million; (c) about 5,800; (d) about 1,000 degrees Kelvin.

25. Solar flares are known to emit bursts of *all but one* of the following: (a) magnetized dust; (b) X-rays; (c) radio waves; (d) energetic protons.

26. Using a partially opaque filter, how can a naked-eye observer determine the period of rotation of the Sun?

27. The rotational period of the Sun can be measured easily by an amateur astronomer who has a small telescope. Describe the procedure.

28. The synodic period of rotation of the equatorial belt of the Sun is (a) 27 days; (b) 365 days; (c) 11 years; (d) $23^h\ 56^m$.

29. The period of the sunspot cycle is about (a) 11 years; (b) 27 days; (c) 176 years; (d) 18.6 years.

30. The diverse radiations from solar flares have been shown to produce *all but one* of the following terrestrial effects: (a) blackouts in radio communications; (b) changes in the orbital motion of the Earth; (c) auroral displays; (d) magnetic storms.

31. The soft X-ray emission of the Sun varies with time by (a) less than 1%; (b) less than 10%; (c) an undetectable amount; (d) factors of ten or more.

32. The solar wind (a) consists of fine dust blown outward from the Sun by radiation pressure; (b) consists of hot ionized gas, mostly hydrogen, moving radially outwards from the Sun's corona; (c) has been searched for in space experiments but has not been detected; (d) is the name given to the fresh breeze that often occurs on the Earth just after sunrise.

☆ ☆ ☆

CHAPTER 2

Motion of the Planets

33. Why are the planets called celestial "wanderers"?

34. Why are the positions of planets not shown in standard star charts?

35. During the course of a year the Sun moves across the star field through the twelve signs of the Zodiac. Name three constellations of the Zodiac.

36. Describe personal observations of one or more planets that you have made during the past six months.

37. During the historical development of an understanding of planetary motion, one of the following persons was the most influential in advocating the heliocentric hypothesis. Which one? (a) Copernicus; (b) Tycho Brahe; (c) Aristotle; (d) Ptolemy.

38. What is the principal conceptual difference between the Ptolemaic and Copernican models of planetary motion?

39. What is the single most significant objection to Ptolemy's model of the solar system?

40. Explain briefly whether or not Ptolemy's model of planetary motion is compatible with Newton's law of gravitation and with his three laws of motion.

41. In their simplest forms the geocentric and heliocentric hypotheses of planetary motion are geometrically equivalent. In what sense is the heliocentric hypothesis superior?

42. Consider two planets of the solar system, Earth and Mars:

    (a) According to the heliocentric hypothesis advocated by Copernicus, the planets Earth and Mars revolve uniformly, but at different rates, in the counterclockwise sense along separate circles of different radii, each centered on the Sun.

    (b) According to the geocentric hypothesis of Hipparchus and Ptolemy, Mars revolves at a uniform rate in the counterclockwise sense along a small circle (the epicycle) whose center, in turn, revolves at a uniform rate in the counterclockwise sense along a larger circle (the deferent) centered on the Earth.

    For the sake of a simple comparison, assume that all motions occur in the same plane. Further

    (c) Under hypothesis (a): Let the orbit of Mars have a radius 1.59 times that of the orbit of the Earth; and let the sidereal period of revolution of the Earth be one year and that of Mars be two years.

    (d) Under hypothesis (b): Let the respective sidereal periods of revolution of Mars along the epicycle be one year and of the center of the epicycle along the deferent be two years and let the respective radii of the epicycle and the deferent be 1.0 and 1.59 units.

    Geometric equivalence of (a) and (b) is demonstrated *if* the radial distances between the two planets and the celestial longitude of Mars as observed from the Earth are the same functions of time under either hypothesis. A graphical solution is suggested, though a more rigorous trigonometric solution may be preferred by some students.

43. Observations of the progression of the phases of Venus (a) support the Ptolemaic hypothesis; (b) support the Copernican hypothesis; (c) do not discriminate between the Ptolemaic and Copernican hypotheses; (d) have never been satisfactorily achieved.

44. What was the principal astronomical contribution of Copernicus?

45. Tycho Brahe's principal contribution to astronomy was (a) his detailed observations of features on the surface of the Moon; (b) his theory of the motions of the planets; (c) his observations of the apparent motion of the planets on the star field; (d) his discovery of sunspots.

46. In the early history of astronomy, some people believed that the Sun revolved about the Earth and others believed that the Earth revolved about the Sun. Which view is "correct" and why?

47. Which one of the following persons was the first to recognize the correct geometric form of the orbits of the planets? (a) Tycho Brahe; (b) Copernicus; (c) Ptolemy; (d) Kepler.

48. Kepler derived his three laws of planetary motion (a) from Newton's laws of motion and of gravitation; (b) following a helpful suggestion by Galileo; (c) from study of Tycho Brahe's observational data; (d) within the general context of the Ptolemaic model of the solar system.

49. Kepler discovered his three laws of planetary motion on the basis of (a) Newton's universal law of gravitation; (b) Tycho Brahe's observations of the apparent motion of planets on the star field; (c) Roemer's observations of the satellites of Jupiter; (d) the observed motion of the Moon about the Earth.

50. State Kepler's three laws of planetary motion.

51. In what sense did Kepler's discovery of his first law favor the Copernican model of the solar system over the Ptolemaic model?

52. Solve the following equation for P when a = 16:

$$P^2 = a^3.$$

53. Describe how to draw an ellipse.

54. Sketch an ellipse and describe its basic geometrical properties.

55. According to Kepler the center of the Sun is at one of the two foci of the elliptical orbit of a planet. What is at the other focus?

56. Define (a) mass; (b) weight; (c) velocity; (d) speed; (e) acceleration.

57. A motorcycle traveling at a constant speed of 30 miles per hour goes West for two hours and then North for two hours. Find its (a) average (scalar) speed for the four-hour period; (b) average (vector) velocity (magnitude, direction, and sense) for the four-hour period; and (c) average (vector) acceleration (magnitude, direction, and sense) for the four-hour period.

58. How many different controls are there in an automobile for causing it to accelerate? Name them and give a brief explanation for each. Recall the proper vector meaning of acceleration.

59. An inertial frame of reference is one in which (a) an object has no inertia; (b) every object is either stationary or moving with constant velocity; (c) Newton's first law of motion is valid; (d) no object can rotate.

60. An inertial reference system is one in which (a) Newton's first law of motion is valid; (b) all objects move at uniform velocity irrespective of the forces exerted on them; (c) all objects are at rest; (d) Newton's first law of motion is invalid.

61. An object is moving away from a fixed point O at a constant speed in a straight line, as referenced to an inertial coordinate system. Work out its path as referred to a uniformly rotating turntable whose plane is parallel to the object's velocity vector and whose center is at O.

62. The gravitational force on an object by another object (a) varies inversely as the third power of the distance between them; (b) becomes greater as they are brought closer to each other; (c) is a purely theoretical concept that has not yet been confirmed observationally; (d) varies directly as the first power of the distance between them.

63. Newton's formula for calculating the magnitude of the gravitational attractive force between two bodies of masses m_1 and m_2 at separation r is

$$\text{(a)} \quad F = G\,\frac{m_1 m_2}{r} \qquad \text{(c)} \quad F = G\,\frac{m_1 + m_2}{r^2}$$

$$\text{(b)} \quad F = G\,\frac{m_1 r^2}{m_2} \qquad \text{(d)} \quad F = G\,\frac{m_1 m_2}{r^2}$$

64. Write the formula for Newton's law of gravitation, define each term, and add any other necessary remarks for making a complete statement of the law.

65. Prove that the gravitational force exerted by a thin, uniform spherical shell on a unit mass (a) at any point P *internal* to the shell is zero and (b) at any point P′ *external* to the shell is the same as though the entire mass of the shell were concentrated at its center. Generalize result (b) to the case of a solid sphere whose mass density is a function only of the radial distance from its center.

66. The quantity G in Newton's law of gravitation (a) has been determined by a laboratory experiment; (b) has been determined by comparing the centripetal acceleration of the Moon in its orbit with the acceleration of a falling object near the Earth; (c) is derived from a comparative study of planetary orbits; (d) is different for different substances (e.g., lead and aluminum).

67. Experiments by Cavendish were noteworthy for their determination of the (a) speed of light; (b) libration of the Moon; (c) period of rotation of the Earth; (d) proportionality constant in Newton's law of gravitation.

68. Consider an asteroid (a minor planet) in a highly elliptical orbit about the Sun. At every moment its acceleration vector is (a) directed toward the Sun; (b) tangent to its orbital path; (c) perpendicular to its orbital path; (d) directed toward the Earth.

69. Kepler's laws were formulated originally to describe the motion of the planets about the Sun. (a) What is the theoretical rationale for applying modified versions of these laws to the motions of satellites (natural and artificial) about planets? (b) What essential modifications are required?

70. The vector acceleration of a satellite of the Earth in an elliptical orbit is (a) constant in magnitude and direction; (b) tangent to the orbit at every point; (c) always directed toward the center of the Earth; (d) constant in magnitude but variable in direction.

71. A planet is moving along a Keplerian ellipse about the Sun. Show by a carefully drawn diagram the direction and sense of its acceleration vector at various positions along its orbit.

72. According to Newton the orbit of any object under the gravitational attraction of the Sun is a conic section. One type of conic section is an ellipse. What are two other types?

73. Describe the characteristics of the following three types of orbits in a central gravitational field: elliptical, parabolic, and hyperbolic.

74. What name is given to the point on a planet's orbit that is closest to the Sun?

75. Kepler's laws of planetary motion are derivable from Newton's three laws of motion and one other basic law. What is the latter?

76. To a high level of accuracy, the orbits of all planets, asteroids, and comets lie in planes passing through the center of the Sun. Explain.

77. Show that it is physically impossible for an unpropelled satellite of the Earth to have an orbit in the form of a halo above the polar region of the Earth.

78. A set of XYZ axes having a fixed orientation with respect to distant stars is carried around the Earth on a spacecraft moving at uniform speed along a circular orbit. Is it an inertial coordinate system? Explain your answer.

79. Newton's law of gravitation (a) provides a simple explanation for the fact that the Earth is rotating; (b) applies to planets (including the Earth) but not to self-luminous bodies (stars); (c) explains the fact that summers on the Earth are warmer than winters; (d) is essential in understanding the motion of the Moon.

80. Newton's law of gravitation implies that (circle *one* or *more* correct answers) (a) the period of revolution of a planet is proportional to the quotient of its mass by the square of its average distance from the Sun; (b) the period of revolution of a planet is approximately independent of its mass; (c) all planets must revolve around the Sun in a counterclockwise sense as viewed from the northern celestial pole; (d) all objects in the solar system affect all other objects.

81. An astronaut (without a tether) steps out of a spacecraft that is orbiting the Earth, but continues to move in an essentially identical orbit. Explain the physical principle involved in understanding this.

82. What are the direction and magnitude of the acceleration of an object moving at constant speed v in a circular orbit of radius r?

83. In newspaper accounts of manned satellite flights it is common to report the speed of the spacecraft. From an astronomer's point of view, the value of the speed is not "news". Explain.

84. Using the fact that the centripetal acceleration of a planet moving at a speed v in a circular orbit of radius r under the gravitational attraction of the Sun is v^2/r, find how v varies with r.

85. What is the physical evidence that the Earth is revolving in an orbit about the Sun and not vice versa?

86. At some point X on the line between the Earth and the Moon the respective gravitational forces on an object by the Earth and by the Moon are equal in magnitude and opposite in direction. Hence, (a) a small object placed at rest at X will remain there forever (ideally); (b) X is the only point on a freely coasting flight between the Earth and the Moon at which an astronaut will be "weightless"; (c) a small object placed at X will initially move away from X at the velocity it has relative to an inertial coordinate system; (d) a spacecraft that passes through X will be torn apart by tidal forces.

87. Why is it impossible for any object to be at rest in the solar system (i.e., with respect to an inertial coordinate system whose origin is at the center of the Sun)?

88. The astronomical unit is defined as the mean distance from the Sun to the Earth *or* as the semimajor axis of the Earth's elliptical orbit about the Sun. Prove that these two definitions are equivalent.

89. The astronomical unit (AU) is a unit of length equal to the (a) mean distance from the Moon to the Earth; (b) diameter of the Sun; (c) circumference of the Earth's orbit; (d) mean distance from the Sun to the Earth.

90. How was the magnitude of the astronomical unit determined before 1960?

91. What is the minimum speed that a spacecraft at 1 AU from the Sun must have in order that it escape from the solar system?

92. The mean orbital speeds and distances from the Sun of three planets are as follows:

	Speed	Distance from Sun
Mercury	48 km s^{-1}	0.39 AU
Earth	30	1.00
Jupiter	13	5.20

Prove that Newton's second law ($F = ma$) is obeyed assuming the validity of the law of gravitational attraction.

93. The sidereal periods of revolution P and the semimajor axes a of the orbits of four satellites of Jupiter are as follows:

Satellite	P (days)	a (arbitrary units)
Io	1.77	0.42
Europa	3.55	0.67
Ganymede	7.15	1.07
Callisto	16.69	1.88

Prove that Kepler's third law is valid for this system.

94. The Earth traverses its orbit about the Sun at nearly constant speed. In a brief statement explain how to reconcile this fact with the Newtonian theory that it is being continuously accelerated toward the Sun.

95. At a speed of 10 km s^{-1} toward the nearest star four light years away, the length of time required for a spacecraft to reach the star is _____ years.

96. The distance to the nearest star (Proxima Centauri) is 270,000 AU. How many years would it take for a spaceship to reach that star if it traveled in a straight line at a uniform speed of 20 km s^{-1}? The orbital speed of the Earth is 30 km s^{-1}.

97. A satellite of the Earth in a close circular orbit has a period of 84 minutes (neglecting atmospheric drag). What is the period of one in an orbit whose perigee and apogee are at 3 and 37 Earth radii, respectively, from the center of the Earth?

98. Earth satellite Explorer 33 had its apogee at a radial distance from the center of the Earth of 70 Earth radii and its perigee at 10 Earth radii. (a) Find its sidereal period of revolution P, given that the period of a satellite in a circular orbit near the Earth is 84 minutes. (b) Find the ratio of its speed at perigee to that at apogee.

99. The University of Iowa's satellite Hawkeye 1 had its perigee at 1.1 Earth radii from the center of the Earth and its apogee at 20.9 Earth radii. (a) What was the semimajor axis of its orbit in the same units? (b) Assuming Kepler's second law, what was the ratio of the speed of the satellite at perigee to its speed at apogee?

100. A satellite is fired from the Earth in such a way that its perigee is at the Earth's surface and its apogee at the Moon's orbit, whose radius is 60 Earth radii. The period of revolution of the satellite is (a) 27.9; (b) 9.9; (c) 4.9; (d) 13.7 days.

101. The mass of the Earth is 81 times that of the Moon and the average distance of the Moon from the Earth is 60 Earth radii. Estimate the maximum radius in Earth radii that the orbit of a satellite of the Moon can have without being "taken away" from the Moon by the Earth. (a) 1; (b) 25; (c) 2; (d) 6.

102. An artificial satellite of the Earth in a low altitude circular orbit has a period of about 1.4 hours. Hence one in a circular orbit with radius of 6.6 Earth radii has a period of _____ hours.

103. A particular satellite of the Earth has the apogee of its orbit at 120 Earth radii and its perigee at 10 Earth radii, both measured from the center of the Earth. By Kepler's second law the ratio of its speed at apogee to that at perigee is (a) $1/12$; (b) 12; (c) $\sqrt{12}$; (d) $1/\sqrt{12}$.

104. A satellite of the Earth in a nearly circular orbit at an altitude of about 200 km has a lifetime in orbit of only a few months. The principal physical cause of its limited lifetime is (a) lunar tidal forces; (b) atmospheric drag; (c) oblateness of the Earth; (d) magnetic drag.

105. As air friction causes a satellite to spiral inward toward the Earth, its orbital speed increases and its period decreases. Explain.

106. What is the shortest period of time during which an unpropelled artificial satellite of the Earth can make one revolution about the Earth?

107. A "synchronous" satellite of the Earth is one whose circular orbit lies in the equatorial plane and whose sidereal period of revolution is 23^h56^m. Given that the period of a satellite whose circular orbit has a radius equal to the radius of the Earth is 84.5 minutes, find the radius of the orbit of a synchronous satellite.

108. A satellite of the Earth is in a circular orbit of radius two Earth radii in the ecliptic plane. Estimate, with the help of a sketch, the fraction of the time that the satellite is in the shadow of the Earth. (a) 50; (b) 8; (c) 16; (d) 32 percent.

109. If the gravitational attraction between the Earth and the Moon were nullified at the time of new Moon without making any other change in the solar system, the Moon would (a) gradually escape from the solar system; (b) strike the Earth; (c) move inward and strike the Sun; (d) follow a slightly eccentric orbit around the Sun similar to that of the Earth.

110. If the mass of the Earth were decreased by a factor of four and its radius decreased by the same factor, the "acceleration due to gravity" on its surface would be _____ g.

111. If the rotation of the Earth were to be stopped by some artificial method, (a) the Moon's orbit would shrink in radius; (b) the Moon's orbit would expand in radius; (c) the Moon's orbit would be unaffected; (d) the Moon would escape from the Earth.

112. Knowing the value of g and the radius of the Earth, assuming the validity of Newton's law of gravitation, and observing the time required for the Moon to make one complete circle around the celestial sphere, one can (a) calculate the average distance from the Earth to the Moon; (b) calculate the mass of the Earth; (c) calculate the mass of the Moon; (d) do none of the above.

113. If the Sun's rotation were stopped, (a) the orbits of planets would be changed markedly; (b) the orbits of planets would remain the same; (c) the pattern of the seasons on Earth would be changed; (d) tides on the Earth would cease.

114. If the mass of the Sun were suddenly increased to twice its present value, what would be the subsequent orbital motion of the Earth?

115. If the mass of the Earth were suddenly reduced to one-half its present value, what would be its subsequent orbit?

116. Consider two objects of different mass revolving in circular orbits of the same radii around the Earth. The orbital speeds of the two objects (a) are in the ratio of their respective masses; (b) are identical; (c) are in the inverse ratio of their respective masses; (d) are in the inverse square ratio of their respective masses.

117. The gravitational perturbations of the several planets on the orbits of others (a) are so large as to make Kepler's laws of very little value; (b) produce effects that have been confirmed by observation; (c) average to zero over a period of a few hundred years; (d) are so small that they have never been observed to exist.

118. An object placed at a particular point X on the line between the Sun and Jupiter experiences equal and opposite gravitational attractions by these two bodies. The mass of the Sun is 1047 times as great as that of Jupiter and the radius of Jupiter's orbit is 5.2 AU. How far is X from Jupiter?

119. Which one of the following statements is *false*? (a) All planets revolve around the Sun in the same sense. (b) Two of the major planets have no known natural satellites. (c) The orbital speed of Pluto is much less than that of Mercury. (d) The rings of Saturn consist of several thin, rigid sheets of matter revolving at different rates around the axis of the planet.

120. An engineer who is calculating how to launch a satellite eastward from Cape Canaveral into a low altitude circular orbit adds in the eastward velocity of the launching site as inferred from the rotation of the Earth. If the engineer neglects to do this but makes an otherwise correct calculation, the satellite will (a) go into a quite elliptical orbit; (b) orbit the Earth from east to west; (c) escape the gravitational field of the Earth; (d) fall back to the Earth at some point less than 180 degrees East of the launch site.

121. A NASA engineer planned the launching of a satellite due East from Cape Canaveral after having calculated launch conditions intended to place the satellite in a low circular orbit about the Earth but the engineer neglected to take account of the rotation of the Earth. What would have been the resulting flight of the satellite?

122. By means of the usual launching technique (one to three stages of short duration thrust by rocket boosters fired in rapid succession) at Cape Canaveral, Florida, it is impossible to inject a satellite into an orbit whose inclination to the equatorial plane of the Earth is less than 28.3 degrees, the latitude of the launching site. Explain.

123. Assuming the correctness of the statement in the preceding problem, describe how it is possible to deliver a satellite into a circular geosynchronous orbit of radius about 6.6 Earth radii in the Earth's equatorial plane, following a launch from Cape Canaveral.

124. A spacecraft is launched from the Earth in such a way that its velocity relative to the Earth (after escape from the Earth's gravitational field) is equal in magnitude and opposite in direction to the orbital velocity of the Earth. Hence, it will fall into the Sun along a straight line. What will be the length of time required to reach the Sun?

125. In shooting a spacecraft from the Earth in order to hit the Sun it must be fired (a) in the direction of the Earth's orbital motion at a net escape speed of 30 km s^{-1}; (b) in the direction perpendicular to the Earth's motion inward toward the Sun at a net escape speed of 30 km s^{-1}; (c) in the direction opposite to the Earth's orbital motion at a net escape speed of 60 km s^{-1}; (d) in the direction opposite to the Earth's orbital motion at a net escape speed of 30 km s^{-1}. The Earth's orbital speed is 30 km s^{-1}.

126. At the surface of the Earth, the "escape speed" from the gravitational field of the Earth is 11.2 km s^{-1}. The resulting (idealized) trajectory is a parabola with respect to the center of the Earth. As viewed in a heliocentric inertial coordinate system, the trajectory remote from the Earth is (a) a parabola; (b) an ellipse; (c) an hyperbola.

127. As a comet or spacecraft flies by a planet, its planetocentric velocity vector is changed in direction but not in magnitude. Analyze the dynamics of such an encounter and show how a net change in heliocentric kinetic energy of the comet or spacecraft can occur despite the fact that the planet's gravitational field is a conservative one. This process is sometimes called "a gravitational assist".

128. Taking the orbits of the Earth and Jupiter to be circular with radii 1 and 5 AU, respectively, and in the same plane, calculate the flight time from the Earth to Jupiter along the minimum energy transfer ellipse.

129. A spacecraft is sent along the minimum-energy transfer ellipse from Jupiter to Saturn. Taking the two orbits to be in the same plane and circular with radii 5 and 10 AU, respectively, calculate the one-way flight time.

130. The semimajor axis of the orbit of Uranus is 19.2 AU. Hence, by Kepler's third law, its sidereal period of revolution is (a) 2.7; (b) 369; (c) 4.4; (d) 84 years.

131. Taking the orbits of the Earth and Mars to be coplanar circles of radii 1.0 and 1.5 AU, respectively, find the time required for an unpowered spacecraft to fly in a Keplerian ellipse from perihelion at the Earth's orbit to aphelion at Mars' orbit.

132. Given that the semimajor axes of the elliptical orbits of the Earth and Saturn are 1 and 10 AU, respectively, calculate the sidereal period of revolution of Saturn.

133. The maximum possible flight time to the Moon along a coasting trajectory from the Earth is about (a) 10.8; (b) 2.6; (c) 4.9; (d) 2.0 days. The radius of the Moon's orbit is 60 Earth radii.

134. The spacecraft Helios I was launched on 10 December 1974 into an orbit about the Sun such that its perihelion distance was 0.3 AU. When did it reach perihelion for the first time?

135. The periodic comet Tuttle 1 has a sidereal period of 13.6 years. What is the semimajor axis of its elliptical orbit about the Sun?

136. The semimajor axis of the orbit of Pluto is 39.44 AU. What is its sidereal period of revolution about the Sun?

137. Estimate the orbital period of a spacecraft whose perihelion is at the Earth's orbit and whose aphelion is at Mars' orbit.

138. The semimajor axis of Saturn's orbit is 9.5 AU. Calculate its sidereal period of revolution about the Sun by comparison with that of the Earth, using Kepler's third law.

139. Pluto's sidereal period of revolution around the Sun is 248 years. What is the semimajor axis of its orbit?

140. Comet Halley has a sidereal period of 76.2 years. What is the semimajor axis of its orbit? Use Kepler's third law.

141. With what subject is Bode's law concerned?

142. List the nine major planets of the Sun, in order outward from the Sun.

143. The semimajor axis of the orbit of Neptune is 30.09 AU. What is its sidereal period of revolution about the Sun?

144. The mass of Jupiter is 1/1047 of that of the Sun. At what radial distance from the planet would a satellite in a circular orbit have a period of revolution of one year?

145. A small satellite of Jupiter (Leda) was discovered in mid-September 1974. The semimajor axis of its orbit is 1.1×10^7 km. The orbit of one of the Galilean satellites, Ganymede, has a period of 7.155 days and a semimajor axis of 1.07×10^6 km. Calculate the period of Leda.

146. The mass of Jupiter is 318 times that of the Earth. At what radial distance from Jupiter would a satellite in a circular orbit have a period of 27.3 days?

147. The sidereal period of revolution of lunar-orbiting satellite Explorer 35 is 11.5 hours. The major axis of its elliptical orbit is 6.9 lunar radii. Find the period of a satellite that is in a circular orbit about the Moon just above its surface.

148. Sketch the positions of several planets at a particular moment and show which way they would move if the Sun suddenly ceased to exist.

149. If the mass of the Sun were suddenly reduced to one-half of its present value, what would be the subsequent motion of the Earth?

150. Suppose that the gravitational attraction of the Sun suddenly ceases to exist. Which of the two planets, the Earth or Neptune, would reach a radial distance of 50 AU from the Sun first? Take the two orbits to be circular with radii of 1.0 and 30 AU, respectively.

151. If the Sun's gravitational force suddenly ceased to exist, how long would it take the Earth to reach a distance of 50 AU from the Sun?

152. Prove that the period of revolution of a small satellite in a close circular orbit about a spherically symmetric planet is independent of the size of the planet and inversely proportional to the square root of the planet's mean density.

153. Explain why a close satellite of any one of the planets and of the Moon has a period of revolution lying in the narrow range 1.4 to 3.9 hours, despite their enormous differences in size and mass.

154. A grain of sand is placed in a close orbit about a steel ball bearing whose mean density is 7.8 g cm^{-3}. Estimate its period of revolution. The system is remote from any other gravitating body.

155. The period of revolution of Mariner 9 in its orbit about Mars is 12 hours. Estimate its semimajor axis in units of the radius of Mars assuming that Mars has a mean density 0.72 that of the Earth.

156. A tunnel is bored diametrically through a solid spherical planet of uniform density ρ, radius a, and mass M. A small object of mass m is released from rest at the mouth of the tunnel on the surface of the planet. Describe the object's subsequent motion.

157. The orbital speed of Mars is 24 km s^{-1} and of the Earth, 30 km s^{-1}; and the radii of their orbits (assumed circular, concentric, and coplanar) are in the ratio 1.52/1.00. Find the maximum value of the time rate of change of the distance between them.

158. The plane of Saturn's rings passed through the Earth three times during 1966 (2 April, 29 October, and 18 December). During what year would one expect the rings to next be edge-on to the Earth?

159. The plane of Saturn's rings passed through the Sun on 15 June 1966. Estimate the next subsequent date on which this situation occurred.

160. Take the orbits of Saturn and Neptune to be circular and in the same plane with respective radii 10 and 30 AU. Further, take their respective periods of revolution around the Sun to be 30 and 165 years. Estimate how many times in 165 years a hypothetical observer on Neptune could see the rings of Saturn edge-on.

161. The mean distance from the Earth to the Sun is called "the astronomical unit" of distance. In terms of this unit, what is the mean distance of Pluto, the most remote known planet, from the Sun?

162. Estimate the greatest angle between the Sun and Mercury that can ever occur (as seen from the Earth).

163. Of the eight other major planets, the one that comes the closest to the Earth is (a) Mercury; (b) Mars; (c) Venus; (d) Jupiter.

164. Describe a geometric method of finding the distance from the Sun to Mars in terms of the astronomical unit.

165. Show how Kepler determined the radius of Mars' orbit.

166. The mass of the Sun is 1,047 times as great as that of Jupiter, whose nearly circular orbit has a radius of 5.2 AU. Find the ratio of the gravitational force of the Sun on the Earth to that of Jupiter on the Earth at opposition.

167. How can a radar be used to determine the absolute value of the astronomical unit?

168. By observing the motion of Saturn relative to the star field over a period of many years, it is found that its average motion is (a) eastward by about 6 degrees per year; (b) westward by about 12 degrees per year; (c) eastward by about 12 degrees per year; (d) westward by about 6 degrees per year.

169. Sketch the configuration of the Sun, the Earth, and Venus when the latter is at greatest elongation East.

170. The planet Venus is always an "evening star" when (a) its right ascension is greater than 12 hours; (b) its right ascension is less than 12 hours; (c) it is at eastern elongation; (d) it is at western elongation.

171. Which one of the following planets can pass in front of the Sun (as viewed from the Earth)? (a) Mercury; (b) Mars; (c) Jupiter; (d) Pluto.

172. An inferior planet is closest to the Earth at (a) superior conjunction; (b) inferior conjunction; (c) greatest elongation West; (d) greatest elongation East.

173. The greatest solar elongation of Venus is about 48 degrees. Therefore, the greatest distance of Venus from the Sun is about _____ AU.

174. At the present date the solar elongation of Venus is about (a) _____ degrees East or (b) _____ degrees West.

175. Sketch the relative positions of the Sun, the Earth, and Saturn when Saturn is in opposition.

176. What is the name of the orbital configuration in which a superior planet is at the least distance from the Earth?

177. If Mars crosses an observer's meridian at local midnight, it is at (a) western quadrature; (b) conjunction; (c) eastern quadrature; (d) opposition.

178. Mars is best observed near the time of its (a) opposition; (b) conjunction; (c) eastern quadrature; (d) western quadrature.

179. Which one of the following planets can never be occulted (i.e., obscured) by the full Moon? (a) Saturn; (b) Venus; (c) Mars; (d) Jupiter.

180. Only one of the following planets can ever pass in front of (i.e., transit) the Sun as viewed from the Earth. Which one? (a) Venus; (b) Mars; (c) Saturn; (d) Pluto.

181. Show with a diagram the reason that Mars appears to an observer on the Earth to move westward (retrograde) with respect to the star field during certain periods of time.

182. Retrograde (westward) motion of a superior planet as viewed relative to the star field occurs near the time at (a) inferior conjunction; (b) western quadrature; (c) opposition; (d) eastern quadrature.

183. Retrograde motion of an inferior planet (Mercury or Venus) as viewed relative to the star field occurs near the time at (a) inferior conjunction; (b) superior conjunction; (c) opposition; (d) greatest elongation East.

184. Retrograde motion of a superior planet (e.g., Mars) occurs near the time of its opposition. What is the configuration of an inferior planet (e.g., Venus) near the time at which its motion is retrograde?

185. The angular diameter of a superior planet as viewed from the Earth is largest when the planet is at (a) eastern quadrature; (b) western quadrature; (c) opposition; (d) conjunction.

186. On 25 November 1976, Mars was in conjunction. It was impractical to receive data from the Viking landers and orbiters at Mars for several weeks around this day. Why?

187. Which one of the following planets can never appear at opposition? (a) Mars; (b) Venus; (c) Jupiter; (d) Saturn.

188. The angular diameter of an inferior planet as viewed from the Earth is largest when the planet is at (a) greatest elongation East; (b) inferior conjunction; (c) superior conjunction; (d) greatest elongation West.

189. The planet Mercury as observed from the Earth never appears at an angle greater than about 28 degrees from the Sun. What is Mercury's greatest distance from the Sun?

190. In observing the apparent motion of Mars on the star field for a period of two years it is found that its *average* motion is (a) northward; (b) eastward; (c) sometimes westward; (d) westward.

191. Sketch the relative positions of the Sun, the Earth, and Venus when Venus is at inferior conjunction.

192. Show by a diagram what is meant by the greatest elongation West of Mercury.

193. Define the synodic period of a planet.

194. The synodic period of Venus is 19.2 months and the radius of its orbit is 0.72 AU. Hence, the interval of time between its greatest elongation East (in the evening sky) and its greatest elongation West (in the morning sky) is about (a) 9.5; (b) 4.7; (c) 14.4; (d) 19 months.

195. The sidereal period of revolution of Venus is 225 days. Find its synodic period.

196. During 1978, the following distinctive orbital configurations of Venus occurred on the dates listed:

Superior conjunction	22 January
Greatest elongation East	29 August
Stationary	18 October
Inferior conjunction	7 November
Stationary	26 November

Hence Venus' synodic period is about _____ days.

197. During 1979–80, the following orbital configurations of Venus occurred on the dates shown:

Greatest elongation West	18 January	1979
Superior conjunction	25 August	1979
Greatest elongation East	5 April	1980
Inferior conjunction	15 June	1980
Greatest elongation West	24 August	1980

What is the synodic period of Venus?

198. The sidereal period of revolution of Mercury is 88 days. Find its synodic period.

199. The synodic period of revolution of Mars is 780 days. What is its sidereal period?

200. The synodic period of Mercury is 116 days. Find its sidereal period.

201. The sidereal period of revolution of Jupiter is 4,333 days. Find its synodic period.

202. What is the sidereal period of revolution of a superior planet that appears in opposition exactly once every two years?

203. The sidereal periods of revolution of the Earth and Mars are 1.0 and 1.88 years. Find the synodic period of Mars if it were revolving about the Sun in the retrograde sense (i.e., clockwise as viewed from the northern ecliptic pole).

204. As observed from Venus what is the interval of time between successive occasions at which the Earth is in a direction opposite to that of the Sun? The sidereal period of revolution of Venus is 225 days.

205. The sidereal periods of revolution of Venus and Mars are 225 and 687 days, respectively. Hence the synodic period of Venus as seen from Mars is (a) 169; (b) 462; (c) 335; (d) 912 days.

206. The *American Ephemeris and Nautical Almanac* for 1975 has the following entries for Mercury.

Greatest elongation West	6 March
Superior conjunction	18 April
Greatest elongation East	17 May
Inferior conjunction	10 June
Greatest elongation West	4 July
Superior conjunction	1 August
Greatest elongation East	13 September

Find the synodic period of Mercury.

207. An abbreviated table of the orbital configurations of Mercury during 1979 is as follows:

Superior conjunction	9 February	and	29 May
Greatest elongation East	8 March	and	3 July
Inferior conjunction	24 March	and	31 July
Greatest elongation West	21 April	and	19 August

Find the synodic period of Mercury.

208. An abbreviated table of the orbital configurations of Mercury during 1983 is as follows:

Greatest elongation East	21 April, 19 August, and 13 December
Inferior conjunction	16 January, 12 May, 15 September, and 31 December
Greatest elongation West	8 February, 8 June, and 1 October
Superior conjunction	26 March, 9 July, and 30 October

Find the synodic period of Mercury.

209. What is the sidereal period of an inferior planet that appears in inferior conjunction exactly once a year?

210. Suppose that a superior planet has a synodic period of two years. It is at opposition on 1 January and at quadrature on 1 May. What is its distance from the Sun in astronomical units? Assume circular orbits. Could an actual planet have such an orbit? Explain.

211. Suppose that a satellite of the Earth revolved about the Earth in the ecliptic plane in the opposite sense to the Moon's motion (i.e., clockwise as viewed from the northern ecliptic pole) and had a sidereal period of 90 days. What would be its synodic period?

212. The sidereal periods of revolution of Jupiter and Saturn are approximately 12 and 30 years, respectively. On a particular date, their respective heliocentric longitudes are equal. How many years will elapse before the next occasion on which their heliocentric longitudes are equal?

213. Suppose that the planets Jupiter, Saturn, Uranus, and Neptune are in four-fold conjunction (i.e., all lie on the same radial line from the Sun) at a particular moment. Calculate the lapse of time before the occurrence of the next approximate four-fold conjunction (to within ± 10 degrees of longitude). Assume the orbits to be circular and in the same plane with respective sidereal periods of 11.862, 29.458, 84.014, and 164.793 years.

☆ ☆ ☆

CHAPTER 3

Sun, Earth, and Moon

214. It is sometimes said that Cavendish's balance experiment is a method for "weighing the Earth". What, exactly, is the physical constant that is measured in this experiment?

215. How can the mass of the Earth be determined?

216. The North-South distance (measured along the surface of the Earth) from New Orleans to Peoria is 770 miles. It is observed on a particular day that the Sun's altitude when it is on the observer's meridian (local noon) is 11 degrees greater at New Orleans than it is at Peoria. What is the radius of the Earth?

217. Describe a simple geometric method for finding the radius of the Earth.

218. The circumference of the Earth is about 40,000 km. Therefore, one minute of latitude corresponds to a North-South distance of _____ km on the surface of the Earth.

219. For an observer at height H above the surface of the sea, the distance d to the horizon is given by the approximate formula

$$d = \sqrt{2HR}$$

where R is the radius of the Earth (= 6,371 km). What is d in kilometers when H = 30 meters? Note that all quantities must be in the same units when doing the calculation.

220. The pilot of an aircraft at 30,000 ft can see a sea-level city as distant as about (a) 420; (b) 67; (c) 106; (d) 210 miles.

221. The peak of a coastal mountain can be just seen on a clear day from a small boat 50 miles away. The approximate height of the peak above sea level is (a) 1,700; (b) 550; (c) 8,000; (d) 3,400 ft.

222. The radii of the Earth and the Sun are 6,371 and 696,000 km, respectively. About how many Earths would fit into the volume occupied by the Sun?

223. The most satisfactory determination of the age of the Earth has come from (a) study of the relative abundances of various radiogenic elements in granites; (b) measuring changes in the rate of rotation of the Earth; (c) classifying fossils in ancient limestone beds; (d) measuring changes in the precise orbit of the Moon; (e) accurate observation of the precession of the equinoxes.

224. The age of the Earth has been determined by radioactive dating to be about (a) 40 billion; (b) 4 billion; (c) 1 billion; (d) 4 million years.

225. The most widely accepted age of the Earth is about (a) 4×10^9 years; (b) 5 million years; (c) 5×10^{10} seconds; (d) 4.5×10^{10} years.

226. On the basis of radioactive dating of samples of terrestrial and lunar material it has been determined that the Earth and the Moon both solidified about (a) 100 million; (b) 4.5 billion; (c) 50 billion; (d) 4.5 million years ago.

227. Consider the radioactive decay of Rb^{87} to Sr^{87} with a half-life of 5×10^{10} years. Estimate the approximate age of a sample in which the present atomic abundance ratio $Rb^{87}/Sr^{87} = 18$. Assume that there was no Sr^{87} in the original sample and that it is not the product of any other radioactive process.

228. The radioactive element Th^{232} decays to the stable element Pb^{208} with a half-life of 14 billion years. In a particular geological sample, the atomic abundance ratio Pb^{208}/Th^{232} is found to be 0.25. Estimate the age of the sample (since solidification) assuming no other source of Pb^{208}.

229. If the mass of the Earth were increased by a factor of four and its radius increased by the same factor, the "acceleration due to gravity" on its surface would be _____ g.

230. The mean density of the Earth is 5.52 g cm^{-3}. What is the significance of this fact in terms of the composition of the Earth?

231. What evidence would you cite to refute a claim that the Earth is made mostly of iron?

232. The departure of the shape of the Earth from that of a sphere is caused primarily by (a) lunar tidal forces; (b) its rotation; (c) its orbital motion; (d) solar tidal forces.

233. The mean temperature of the surface of the Earth is about 270 degrees Kelvin. What would this temperature be if the Earth were at 16 AU from the Sun? The surface temperature of an inert body varies as the inverse square root of its distance from the Sun.

234. The temperature at the center of the Earth is estimated to be about 6,400 degrees Kelvin. What are the two principal causes of such a high temperature?

235. Cite some common knowledge that proves that the interior of the Earth is very hot.

236. Why do stars "twinkle"?

237. The blue color of the sky is attributable to which of the following optical phenomena: (a) scattering; (b) refraction; (c) diffraction; (d) reflection; (e) interference?

238. The blue color of the sky on a clear day is caused primarily by (a) sunlight scattered in the atmosphere; (b) atmospheric refraction of sunlight; (c) sunlight reflected from the surface of the Earth; (d) atmospheric scintillation.

239. Because of the atmosphere of the Earth, the observed altitude of an astronomical object always appears to be greater than its true geometrical value. This optical effect is called (a) diffraction; (b) parallax; (c) refraction; (d) absorption.

240. Atmospheric refraction causes (a) the apparent altitude of a star to be less than its true altitude; (b) the apparent position of a star to be to the left of its true position; (c) the apparent position of a star to be to the right of its true position; (d) the apparent altitude of a star to be greater than its true altitude.

241. The density of the Earth's atmosphere decreases by a factor of two for every increase of altitude of about 5.5 km. Hence the density of the atmosphere at the top of Mt. Everest, 29,000 ft (8,840 meters) above sea level, is less than that at sea level by a factor of about (a) 9.5; (b) 2.1; (c) 3.0; (d) 5.1.

242. If the Earth were cooled so that the oceans froze and the atmosphere became liquid air, about how thick a layer of liquid air would cover the surface of the Earth?

243. The most abundant constituent of the Earth's atmosphere is (a) oxygen; (b) carbon dioxide; (c) water vapor; (d) nitrogen.

244. Ozone is a minor, natural constituent of the Earth's upper atmosphere. (a) What is ozone? (b) At about what height is its density a maximum? (c) What is its importance to humans?

245. Discuss whether or not it is possible, in principle, for an aircraft to fly around the Earth in the equatorial plane in less than 84.5 minutes.

246. In order to observe the effects of "weightlessness" (i.e., free fall), a tall evacuated tower is erected on the Earth. How tall must it be to provide free fall of 10 seconds duration?

247. The general magnetic field of the Earth is caused primarily by (a) electrical currents in its ionosphere; (b) a large, magnetized iron core; (c) electrical currents in its interior; (d) iron ore deposits in its outer crust.

248. The basic cause of aurorae ("northern and southern lights") in the Earth's atmosphere is (a) soft x rays from the Sun; (b) the solar wind; (c) intermittent bursts of solar radio noise; (d) reflection of sunlight from snow and ice in the arctic and antarctic regions of the Earth's surface.

249. The radiation belts of the Earth (a) would not exist if the Earth were not magnetized; (b) are composed principally of radioactive nuclei; (c) are composed principally of neutral hydrogen gas; (d) have their greatest particle populations above the northern and southern magnetic poles of the Earth.

250. Describe the motion of an electrically charged particle in the external magnetic field of a magnetized planet.

251. What relationship of the axis of rotation of the Earth to the plane of its orbit would result in no change of seasons during the course of a year?

252. The full Moon is eleven times as bright as the quarter Moon. One might expect it to be only twice as bright. Suggest an explanation.

253. It might be expected that the sum of the light from the first quarter Moon and the light from the last quarter Moon would be equal to the light from the full Moon. Actually it is less than 20 percent. Suggest a line of thought by which this fact may be understood.

254. The Moon (a) always rises in the west and sets in the east; (b) rises about 50 minutes later each night; (c) rises about 50 minutes earlier each night; (d) is full when it crosses the meridian at noon local time.

255. The Moon (a) rises about fifty minutes earlier every night; (b) is full when it crosses the meridian at noon local time; (c) is full when it crosses the meridian at sunset; (d) is full when it crosses the meridian at midnight local time.

256. The point on the Moon's orbit that is the closest to the center of the Earth is called the (a) perihelion; (b) apogee; (c) aphelion; (d) perigee.

257. If the Moon moves about 13 degrees eastward on the star field in a 24-hour period, its sidereal period is how many days?

258. The Moon's apparent motion on the star field during the course of a month is approximately along the (a) celestial equator; (b) prime meridian; (c) tropic of Capricorn; (d) ecliptic.

259. The point on the Moon's orbit that is closest to the Earth is called the (a) ascending node; (b) descending node; (c) perigee; (d) apogee.

260. Select the most nearly correct answer concerning the apparent motion of the Moon relative to the star field: (a) westward along the ecliptic at about 13 degrees day^{-1}; (b) eastward along the ecliptic at about 13 degrees day^{-1}; (c) principally northward; (d) westward along the celestial equator at about 13 degrees day^{-1}.

261. An object consisting of two equal masses m connected by a rigid, massless rod of length ℓ is placed in orbit around the Earth. Prove that the equilibrium orientation of the object is with the rod lying along a radial line through the center of the Earth and find the period of its oscillation about that line. Compare this case with that of the Moon.

262. The length of time between two successive new Moons is called the (a) sidereal; (b) nodal; (c) synodic; (d) Julian month.

263. The sidereal period of revolution of the Moon about the Earth is 27.32 days. What is its synodic period?

264. Approximately the same face of the Moon is always toward the Earth. The period of time between successive full Moons is 29.53 days. Hence, referenced to an inertial coordinate system, the Moon (a) is rotating counterclockwise (as viewed from the northern celestial pole) with a period of 27.32 days; (b) is rotating clockwise with a period of 29.53 days; (c) is rotating counterclockwise with a period of 29.53 days; (d) is not rotating.

265. The sidereal period of rotation of the Moon is equal to (a) its sidereal period of revolution; (b) its synodic period of revolution; (c) one sidereal year; (d) one tropical year.

266. When the Moon is in opposition (relative to the Sun) its phase is (a) full; (b) new; (c) first quarter; (d) last quarter.

267. As referenced to a star chart the Moon (a) always has the same phase when it is at the same right ascension; (b) never crosses the ecliptic; (c) moves mainly toward the West; (d) moves mainly toward the East.

268. How can you find the sidereal period of revolution of the Moon by your own observations?

269. Describe a simple observational program for finding the approximate inclination of the Moon's orbit to the ecliptic.

270. The phase of the Moon is about the same whenever (a) it is at perigee; (b) its right ascension is the same; (c) its right ascension differs by the same amount from the right ascension of the Sun; (d) it is at the ascending node of its orbit on the ecliptic.

271. Moonlight, at full Moon, is $1/400,000^{th}$ as bright as sunlight. If the whole sky were covered with full Moons, how would the illumination on the Earth's surface compare with that of sunlight? Angular diameter of the Moon = 0.5 degree.

272. An observer on the Earth finds that the diameter of the Moon subtends an angle of 0.5 degree. Hence the Moon's distance is greater than its diameter by a factor of about (a) 115; (b) 20; (c) 720; (d) 57.

273. At what time of the year does a first quarter Moon have a right ascension of about 0^h?

274. Full Moon (a) always occurs at the ascending node of its orbit on the ecliptic; (b) usually occurs within a day of perigee passage; (c) sometimes occurs near the descending node; (d) never occurs near apogee.

275. Name at least one month of the year during which the full Moon is below the horizon of an observer at the northern pole of the Earth.

276. At what time of the year, on the average, does the full Moon have its greatest northerly declination and hence its greatest altitude at meridian passage? (a) winter solstice; (b) vernal equinox; (c) summer solstice; (d) autumnal equinox.

277. If the Moon were substituted for the Sun as an astronomical clock, how many lunar days would occur in a conventional (solar) year?

278. If the Moon revolved around the Earth in the same sense that it now does but with a sidereal period of 91 days, its synodic period (interval between successive new Moons) would be about (a) 121; (b) 182; (c) 73; (d) 274 days.

279. Even if the Moon reflected no light (i.e., were perfectly black), its presence would be revealed by a number of observable effects. Name at least three.

280. The Moon is said to be a triaxial body. What does this mean?

281. The classical method for determining the ratio of the mass of the Moon to the mass of the Earth depends upon observing (a) the apparent monthly wavering of the orbits of asteroids passing nearby; (b) the parallax of the Moon as viewed against the star field from different points on the Earth's surface; (c) the difference between the sidereal and synodic months; (d) the eccentricity of the Moon's orbit.

282. The orbit of the Moon is inclined at 5 degrees to the ecliptic. What are the maximum and minimum ranges of the Moon's declination during a period of 18.6 years?

283. The mouth of the Mississippi River is about 6,150 meters farther from the center of the Earth than is its source at Lake Itasca, in northern Minnesota. It appears that water flows uphill, contrary to ordinary experience. What is the explanation of this paradox?

284. During a particular year the descending node of the Moon's orbit on the ecliptic was at the autumnal equinox and full Moon occurred on 21 December. What was the altitude of the Moon at meridian crossing on that date as viewed from a station at latitude 42 degrees North?

285. The Moon crosses an observer's meridian at 21^h 34^m Local Mean Time on 11 November. At about what time will it cross the meridian on 15 November?

286. The first quarter Moon crosses the meridian at about (a) 12^h; (b) 18^h; (c) 24^h; (d) 06^h local apparent solar time.

287. The full Moon crosses an observer's meridian at about (a) noon; (b) sunset; (c) sunrise; (d) midnight.

288. At two days after it is new, the Moon can be seen best (a) at about midnight; (b) in the eastern sky about an hour before sunrise; (c) in the western sky about an hour before sunrise; (d) in the western sky about an hour after sunset.

289. On the 23rd of September, the right ascension of the Sun is 12^h. What is the approximate right ascension of a last quarter Moon that occurs on that date?

290. The Moon is due south at about midnight standard time (01^h daylight time). What is the phase of the Moon?

291. What is the phase of the Moon when it is at quadrature?

292. Prove that the terminator of the Moon is a semi-ellipse whose major axis joins the two tips of the terminator and is equal in length to the lunar diameter.

293. The last quarter Moon crosses an observer's meridian at a local time of about (a) 18^h; (b) noon; (c) midnight; (d) 06^h.

294. Consider how the Moon might have contributed to the evolution of life on the Earth.

295. An artist depicts the crescent Moon, setting in the evening twilight sky, as shown. Is this drawing approximately correct or not? Explain your answer.

296. The first quarter Moon rises at about the following local time: (a) 06^h; (b) midnight; (c) 18^h; (d) noon.

297. At about what time of day does (a) the last quarter Moon rise? (b) the full Moon set?

298. At about what time is the full Moon on the observer's meridian?

299. On the night of 3–4 November 1971, the Moon, about one and a half days after full phase, passed through the Pleiades. The right ascension of the Sun was 14^h 30^m at that time. Hence, the right ascension of the Pleiades is about (a) 01^h 15^m; (b) 03^h 45^m; (c) 05^h 45^m; (d) 15^h 45^m.

300. The three-day old Moon (waxing crescent) is best observed (a) at about midnight; (b) soon after sunrise; (c) in the early evening after sunset; (d) in the early morning before sunrise.

301. The last quarter Moon rises at about the following local time (a) 06^h; (b) midnight; (c) 18^h; (d) noon.

302. Assess the validity of the following statement: "As I walked out of the library at midnight, I noted the full Moon rising in the East."

303. A northern hemisphere observer reported seeing the full Moon rise at about local midnight. The observer (a) could have been correct if near the equator; (b) could have been correct if on the Tropic of Capricorn; (c) was mistaken, irrespective of location and the time of year; (d) could have been correct if in the arctic.

304. An observer at the center of the "far side" of the Moon (a) sees the Sun rise in the West and set in the East; (b) never sees the Earth; (c) never sees the Sun; (d) is always in sunlight.

305. Suppose that, for a terrestrial observer, the mean lunar day is defined in the same manner as is the mean solar day. In terms of ordinary hours and minutes, what is the length of the mean lunar day? (a) $24^h\ 10^m$; (b) $23^h\ 10^m$; (c) $24^h\ 50^m$; (d) $23^h\ 56^m$.

306. An astronaut on the Moon observes the phase of the Earth to be waning gibbous. Hence the phase of the Moon as seen from the Earth is (a) waxing gibbous; (b) waning crescent; (c) waxing crescent; (d) waning gibbous.

307. An astronaut at the center of the Moon's face (as seen from the Earth) observes that the Sun is on the eastern horizon. Therefore the phase of the Moon is (a) full; (b) first quarter; (c) new; (d) last quarter.

308. At what phases of the Moon would an observer on the Moon see a "half-Earth" (i.e., half of the Earth illuminated and half dark)?

309. An astronaut at the center of the Moon's face (terrestrial subpoint) observes that the Earth is at the zenith and is shining at full phase. Which one or ones of the following statements are true? (a) The Moon is new. (b) The astronaut's selenographic longitude is 180 degrees. (c) The astronaut's selenographic latitude is between 7 degrees North and 7 degrees South. (d) The astronaut will not be able to see the Sun for about seven days. (e) The Moon is full.

310. An astronaut standing on the northern pole of the Moon and looking toward the Earth sees the right-hand half of the Earth sunlit and the left-hand half dark. What is the phase of the Moon as viewed from the Earth?

311. An astronaut standing on the southern pole of the Moon and looking toward the Earth observes that the right half of the Earth is sunlit and the left half dark. Hence the phase of the Moon is (a) first quarter; (b) last quarter; (c) new; (d) full.

312. An astronaut standing on the northern pole of the Moon and looking toward the Earth notes that the Earth is in the waxing crescent phase. What is the phase of the Moon as seen from the Earth?

313. An astronaut at the center of the Moon's face (as seen from the Earth) observes the Sun on the western horizon. Therefore, the phase of the Moon (in Earth observer's terms) is (a) last quarter; (b) full; (c) new; (d) first quarter.

314. A photograph of the Moon shows one portion dark and the remainder bright. How can one distinguish between a phase of the monthly cycle and a partial eclipse?

315. Even during a total eclipse of the Moon by the Earth, the Moon can be seen faintly illuminated. Explain.

316. Suppose that the Moon were a thin circular disk (rather than a sphere) that continuously faced the Earth. Describe how the Moon would appear to an Earth observer at the following phases: (a) new Moon; (b) 5 days after new Moon; (c) 9 days after new Moon; (d) 14 days after new Moon.

317. In terms of Earth days (24 hours), what is the length of the "day" on the Moon, i.e., the interval of time between successive meridian transits of the Sun as observed by an astronaut on the Moon?

318. Describe the phases of the Earth as they would appear to an observer on the Moon.

319. An infrequent event is said to occur "once in a blue Moon". A second full Moon during a particular calendar month is called a blue Moon. How often does a blue Moon occur?

320. The "back" side of the Moon (a) has never been seen by a human observer; (b) is never illuminated by the Sun; (c) has fewer and smaller maria than does the "front" side; (d) is visible from the Earth only by "earthshine" during new Moon.

321. An astronaut en route to the Moon observed that its diameter of 3,500 km subtended an angle of 5.7 degrees. Therefore, the astronaut's distance from the Moon was about (a) 350,000; (b) 35,000; (c) 70,000; (d) 19,950 km.

322. The classical method for determining the distance to the Moon employs the principle of (a) Doppler shift; (b) diffraction; (c) parallax; (d) aberration.

323. The classical determination of the Earth-Moon distance is based on the geometric principle called _____.

324. Explain a simple geometric method for finding the distance to the Moon.

325. The Earth-Moon distance is sixty times as great as the radius of the Earth. The Moon appears to an observer at the northern pole to be further South on the star field than it does to an observer at the southern pole. By about how many degrees? (a) 0.067; (b) 0.96; (c) 0.033; (d) 1.91.

326. About how far from the Earth is the Moon?

327. The distance from the Earth to the Moon is 3.8×10^5 km and the radius of the Earth is 6,400 km. The ratio of the former to the latter is approximately (a) 5.9×10^2; (b) 2.4×10^9; (c) 5.9×10^7; (d) 59.

328. Compute the ratio of the gravitational force of the Sun on the Earth to that of the Moon on the Earth. The Sun is 330,000 times more massive than the Earth and the Moon is 81 times less massive. The Sun-Earth distance is 390 times the Moon-Earth distance.

329. As referred to a sidereal coordinate system with origin at the center of the Earth, the orbital speed of the Moon is approximately (a) 11.2; (b) 7.8; (c) 1.0; (d) 2.4 km s^{-1}.

330. Suppose that the mass of the Moon were $1/200^{th}$ that of the Earth and its radius were $1/10^{th}$ that of the Earth. Then the acceleration of a freely falling object near the Moon's surface would be (a) $1/20^{th}$; (b) one-half; (c) $1/10^{th}$; (d) $1/6^{th}$ that near the Earth's surface.

331. The physical diameter of the Moon is about one-fourth of that of the Earth and its angular diameter as viewed from the Earth is about 0.5 degree. Therefore, the angular diameter of the Earth as seen from the Moon is about (a) 0.125; (b) 4; (c) 2; (d) 5.73 degrees.

332. The Moon's sidereal period of revolution is 27.3 days and the mean radius of its orbit is 384,404 km. Hence, its linear orbital speed relative to a sidereal reference system with origin at the center of the Earth is (a) 1,408 km day^{-1}; (b) 1.02 km s^{-1}; (c) 0.163 km s^{-1}; (d) 14,080 km day^{-1}.

333. The diameter of the Moon subtends an angle of 0.5 degree as viewed from the Earth, and the Moon's distance from the Earth is 384,000 km. What is the Moon's diameter?

334. Which one of the following planets can never be occulted (i.e., obscured) by the full Moon? (a) Jupiter; (b) Saturn; (c) Venus; (d) Mars.

335. What is meant by libration of the Moon?

336. If the Moon had a mass equal to that of the Earth but its present size, the gravitational acceleration at its surface would be about (a) 26; (b) 4; (c) 2; (d) 13 g.

337. One of the principal scientific results of the Apollo program was finding that the age of certain surface samples is about 4.5 billion years. Briefly, what is the principle involved in such a determination?

338. To an observer on the Moon the angular diameter of the Earth is 1.9 degrees. Hence the Moon's distance is greater than the diameter of the Earth by a factor of about (a) 720; (b) 115; (c) 30; (d) 60.

339. In early 1971 the phases of the Moon were as follows:

19 January	Last Quarter
26 January	New
2 February	First Quarter
10 February	Full

Suppose that an astronaut landed at the center of the Moon's face on 2 February. (a) Find the approximate altitude and azimuth of the Sun at that site. (b) Describe the appearance (phase) of the Earth as seen by the astronaut.

340. In early 1973 the phases of the Moon were as follows:

4 January	New
12 January	First Quarter
18 January	Full
26 January	Last Quarter

Suppose that on 15 January 1973 an astronaut landed at a point on the center of the Moon's face, as viewed from the Earth. At what altitude was the Sun at the landing site?

341. The temperature of the Moon's surface at the center of its visible disc varies from +260 degrees Fahrenheit at full Moon to −240 degrees Fahrenheit at new Moon. (a) Suggest some practical measures (not including artificially supplied power) for keeping a station there at a reasonably comfortable temperature for human occupancy. (b) At what phases of the Moon would it be most practical, from a thermal point of view, for an astronaut to emerge from the station?

342. At what phases of the Moon should an Apollo flight be launched in order that the temperature conditions at the center of the Moon's disc (as seen from the Earth) be favorable for a landing there?

343. The radius of the Earth is 3.67 times that of the Moon and its mass is 81.3 times that of the Moon. Hence, the gravitational acceleration at the Moon's surface is _____ g.

344. The acceleration of a freely falling object near the Moon's surface is (a) $1/3^{rd}$; (b) six times; (c) $1/16^{th}$; (d) $1/6^{th}$ of that near the Earth's surface.

345. If you can drive a golf ball 100 yards on the Earth, how far would you be able to drive it on the Moon?

346. The study of radioactivity in recovered lunar samples is of little or no use for (a) assessing the bombardment of the Moon by high energy particles; (b) determining the length of time since the Moon solidified; (c) determining the internal temperature of the Moon; (d) studying the chemical constitution of the lunar material.

347. What is the evidence that the Moon's interior was at one time very hot?

348. It is probable that most of the craters on the Moon's surface have been caused by (a) meteoric impacts; (b) volcanoes; (c) moonquakes; (d) the collapse of swampy areas.

349. Most of the craters on the surfaces of Mercury and the Moon are believed to have been caused by (a) volcanoes (now extinct); (b) thermal expansion; (c) meteoric impacts; (d) glaciers.

350. The much smaller number of known meteoric craters on the Earth than on the Moon is believed attributable to (a) shielding of the Earth by the Moon; (b) the greater age of the Moon; (c) the obscuring effects of glaciation, erosion, vegetation, etc. on the Earth; (d) the atmosphere of the Earth.

351. The surface material of maria on the Moon is chemically and structurally similar to the following terrestrial material: (a) granite; (b) limestone; (c) iron ore; (d) basalt.

352. Sketch the basic geometric principle by means of which the height of lunar mountains is determined from photographs of the Moon.

353. In landing equipment gently on the Moon (a) a parachute is helpful but not sufficient to reduce the landing speed to a tolerable value; (b) exclusive reliance is placed on retrorockets to reduce the landing speed; (c) a combination of parachutes and retrorockets is commonly used; (d) the approach speed is reduced by powerful laser beams reflected from the lunar surface.

354. In landing equipment gently on the Moon (a) retrorockets are useless because of the absence of an atmosphere; (b) the approach speed is reduced by using one or more parachutes; (c) reflected radio signals from a powerful transmitter are used to reduce the approach speed; (d) retrorockets are effective despite the absence of an atmosphere.

355. Study of the radioactivity in samples of lunar surface material shows that the age of the Moon is (a) similar to; (b) about three times; (c) about one-third; (d) many times less than that of the Earth.

356. Describe the observational evidence for statements that the Moon has little or no atmosphere.

357. Briefly, explain why the Moon has little or no gaseous atmosphere, and state whether it would be more or even less likely to have an atmosphere if it were at a distance of 10 AU from the Sun.

358. Why does the Moon have essentially no atmosphere?

359. The principal reason that there is virtually no atmospheric gas on the Moon is that (a) it is pulled off by the Earth's tidal forces; (b) all such gas is frozen at the lunar poles; (c) the Moon's gravitational field is too weak to prevent its escape into space; (d) it all collects on the back side.

360. A mixture of the following gases is released from a spacecraft on the surface of the Moon: hydrogen, helium, nitrogen, neon, and xenon. (a) Which gas would be retained for the longest time in the Moon's atmosphere? (b) Why?

361. Concerning the existence of life elsewhere in the universe only one of the following statements is true. Which one? (a) Samples of lunar material brought back by the Apollo astronauts contain fossil remains of small animals. (b) There is no conclusive evidence against it. (c) Living bacteria have been found on the surface of Mars. (d) The red spot on Jupiter has been shown to contain protein molecules.

362. Why do the highest tides (spring tides) occur at times of new and full Moon and the lowest tides (neap tides) occur at the times of first and last quarter Moon?

363. The fact that the Moon rotates so as to keep the same face continuously toward the Earth is attributed to (a) tidal torque; (b) magnetic forces; (c) sunlight; (d) coincidence.

364. Explain why one face of the Moon is always (approximately) toward the Earth.

365. Spring tides on the Earth occur (a) only near times of new Moon; (b) near the times of both full Moon and new Moon; (c) near the times of the first quarter Moon; (d) near the times of the last quarter Moon.

366. List some terrestrial and astronomical effects of tidal forces in the Earth-Sun-Moon system.

367. The average time interval between successive high tides on the Earth is 12^h 25^m. Explain in a few words why this fact alone shows that the Moon is the principal cause of the tides.

368. All *but one* of the following are attributed to tidal forces and/or tidal torques: (a) Kepler's laws of planetary motion; (b) precession of the rotational axis of the Earth; (c) the fact that the same face of the Moon is always toward the Earth; (d) the rings of Saturn.

369. High tide in the New York harbor occurred at 21^h 30^m on 9 October. On 12 October a high tide can be anticipated at (a) 11^h 35^m; (b) 23^h 10^m; (c) 14^h 35^m; (d) 09^h 05^m.

370. The mean interval between high tides on the Earth is (a) 23^h 56^m; (b) 24^h 50^m; (c) 11^h 35^m; (d) 12^h 25^m.

371. High tide in the Boston harbor occurred at 01^h 30^m on 1 November. On 5 November a high tide can be anticipated at (a) 05^h 40^m; (b) 03^h 10^m; (c) 17^h 15^m; (d) 15^h 35^m.

372. Which of the following astronomical events is most likely to cause an earthquake: (a) perigee passage of the new or full Moon; (b) alignment of the outer planets (at approximately the same right ascension); (c) a solar flare; (d) an inferior conjunction of Venus?

373. What is meant by the statement that "the Moon is at the ascending node of its orbit"?

374. What is the relative tide-raising effectiveness of the Sun to that of the Moon? For this approximate calculation assume that the Earth is 80 times more massive than the Moon, and 300,000 times less massive than the Sun and that the Sun is 400 times more distant than the Moon.

375. Sketch the geometric relationship of the Sun, Moon, and Earth at the time of a total eclipse of the Sun.

376. At what phase must the Moon be in order that it be eclipsed by the Earth?

377. A total eclipse of the Sun by the Moon occurs only at (a) first quarter Moon; (b) last quarter Moon; (c) full Moon; (d) new Moon.

378. In order that the Moon eclipse the Sun, it must be (a) near the perigee of its orbit; (b) near the line of intersection of its orbital plane with the Earth's equatorial plane; (c) near the line of intersection of its orbital plane with the ecliptic plane; (d) near the apogee of its orbit.

379. What is meant by the term "eclipse season" in the Sun-Moon-Earth system and what geometrical conditions are necessary for its occurrence?

380. A total eclipse of the Moon is observable (a) only briefly within a narrow stripe across the Earth's surface; (b) from about one-half of the surface of the Earth; (c) only near the time of new Moon; (d) only near the midnight meridian.

381. A total eclipse of the Moon would occur every synodic month if, everything else remaining the same, (a) the orbit of the Moon were a circle with radius equal to its present semimajor axis; (b) the Moon's diameter were twice as great as it is now; (c) the orbit of the Moon were inclined at zero degrees to the ecliptic; (d) the Sun were a point source of light.

382. The length of the umbra of the Earth's shadow is (a) several times greater than the Earth-Moon distance; (b) much less than the Earth-Moon distance; (c) approximately equal to the Earth-Moon distance; (d) about 100 times greater than the Earth-Moon distance.

383. During a total eclipse of the Sun by the Moon the center of totality sweeps across the Earth principally from (a) West to East; (b) North to South; (c) South to North; (d) East to West.

384. Identify the circumstances under which this view of the Moon occurred (black = dark). The Moon was
    (a) at its waxing crescent phase;
    (b) about to enter total eclipse by the Earth;
    (c) emerging from total eclipse by the Earth;
    (d) at its waning crescent phase.

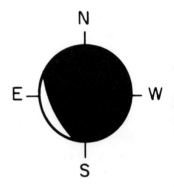

385. If one denotes the Earth-Moon distance by d, it is found that the length of the umbra of the Moon's shadow is (a) approximately equal to d; (b) much less than d; (c) several times greater than d; (d) about sixty times greater than d.

386. A total solar eclipse has its greatest duration when (a) the Sun is at its nearest distance and the Moon at its nearest; (b) the Sun is at its farthest distance and the Moon at its nearest; (c) the Sun is at its nearest distance and the Moon at its farthest; (d) the Sun is at its farthest distance and the Moon at its farthest.

387. Calculate the length of the umbral shadow of a solid sphere of radius one-tenth that of the Sun if it is located 2 AU from the Sun.

388. During a particular year the ascending node of the Moon's orbit on the ecliptic is at right ascension 06^h. Hence, the center of an eclipse season occurs during one of the following months: (a) August; (b) October; (c) March; (d) June.

389. During 1968 the following eclipses occurred:

28–29 March	Partial eclipse of the Sun
13 April	Total eclipse of the Moon
22 September	Total eclipse of the Sun
6 October	Total eclipse of the Moon

Predict the dates of the centers of the two eclipse seasons in 1969. The nodes of the Moon's orbit on the ecliptic regress with a period of 18.6 years so that the average interval between eclipse seasons is about 173 days.

390. The following eclipses occurred during 1969:

18 March	Annular eclipse of the Sun
2 April	Penumbral eclipse of the Moon
27 August	Penumbral eclipse of the Moon
11 September	Annular eclipse of the Sun
25 September	Penumbral eclipse of the Moon

Find the approximate dates of the mid-points of eclipse seasons in 1974.

391. Early in 1971 the following eclipses occurred:

| 10 February | Total eclipse of the Moon |
| 25 February | Partial eclipse of the Sun |

Were there any further eclipses of either the Sun or the Moon during 1971? If so, about when?

392. In mid-1973, the ascending node of the Moon's orbit on the ecliptic plane was at right ascension $18^h 30^m$. Hence the mid-time of an eclipse season during 1973 was during one of the following months: (a) April; (b) November; (c) September; (d) June.

393. During 1974 the following eclipses occurred:

4–5 June	Partial eclipse of the Moon
20 June	Total eclipse of the Sun
29 November	Total eclipse of the Moon
13 December	Partial eclipse of the Sun

    Predict the approximate dates of the centers of eclipse seasons in 1975.

394. The line of nodes of the Moon's orbit on the ecliptic rotates clockwise (as seen from the northern ecliptic pole) through 360 degrees in 18.6 years. Hence the interval of time between two successive eclipse seasons is about (a) 173; (b) 346; (c) 183; (d) 19 days.

395. In early August 1976, the ascending node of the Moon's orbit on the ecliptic plane was at right ascension $14^h 30^m$. Hence, the center of an eclipse season occurred in (a) late July; (b) mid September; (c) early December; (d) late October.

396. If the right ascension of the ascending node of the Moon's orbit on the ecliptic is 06^h, on about what dates of the year will the mid-times of the eclipse seasons occur?

397. During a particular year the right ascension of the ascending node of the Moon's orbit on the ecliptic is 12^h. Hence the center of an eclipse season will occur during (a) May; (b) November; (c) July; (d) March.

398. What is the most difficult feature in the prediction of eclipses many years into the future?

399. In mid-1977 the right ascension of the ascending node of the Moon's orbit on the ecliptic was $13^h 20^m$. At about what date did the center of the first eclipse season of 1977 occur?

400. An interval of time between centers of successive eclipse seasons is most nearly equal to one of the following: (a) 12 months; (b) 11 months; (c) 6 months; (d) 18.6 years.

401. Why is the eclipse year (347 days) less than the tropical year (365 days)?

402. By means of historical records of solar eclipses 2,000 years ago and the Keplerian/Newtonian theory of the orbital motions of the Earth and the Moon, it is found that modern solar eclipses occur three hours earlier than they would if the rotational period P of the Earth had been constant. Assuming that this discrepancy is caused solely by a uniform rate of change of the rotational period of the Earth, find this rate of change in seconds per year.

403. Specify the circumstances under which an annular eclipse of the Sun occurs.

404. Show how to estimate the duration of a total eclipse of the Sun at a particular point on the path of totality.

405. Find by rough calculation the speed of the umbral shadow across the (rotating) Earth during a total eclipse of the Sun. Assume that the shadow moves along the Earth's equator.

406. During a particular total eclipse of the Sun the path of totality lies along the Earth's equator. As seen by an observer on the (rotating) Earth, which way does the spot of totality move and with what speed? Explain the basis for your estimates.

407. During a particular total eclipse of the Sun, the center of the path of totality is along the Earth's equator and the diameter of the umbra of the Moon's shadow at the Earth's surface is 200 km. Calculate the duration of totality for an equatorial observer. Do not neglect the rotation of the Earth.

408. Assuming that the Earth experiences a total eclipse of the Sun on the average every other year, that the path of totality is 80 km wide, and that the path is equally likely to intersect the axis of the Earth within equal increments of distance along its North-South diameter, one expects that at a given position on the Earth's surface a total eclipse will be observable about once in (a) 120; (b) 320; (c) 160; (d) 640 years.

409. Why is the sky not totally dark during a total eclipse of the Sun?

410. A popular hypothesis is that unidentified flying objects (UFO's) are spacecraft carrying intelligent extraterrestrial beings. Suggest several tests of the validity of this hypothesis.

☆ ☆ ☆

CHAPTER 4

The Sky as Observed from the Rotating, Revolving Earth

411. Name several common uses for *The Astronomical Almanac*.

412. The "trace" of the Sun on a star chart during the course of a year is called (a) the celestial equator; (b) the vernal equinox; (c) the galactic equator; (d) the ecliptic.

413. The vernal equinox is the point in the star field occupied by the Sun on about (a) 21 September; (b) 21 March; (c) 21 June; (d) 21 April.

414. What is a stellar constellation?

415. What is the collective name of the twelve constellations through which the Sun passes during a year (as of 2,000 years ago)?

416. Of the 88 constellations, 12 are called "signs of the Zodiac". Why were 12 so designated?

417. What is the astronomical meaning of the term "sign of the Zodiac"?

418. There are ____ "signs" of the Zodiac.

419. Why is Polaris one of the favorite stars of navigators and survey-ors?

420. Which one of the following constellations is *not* one of the "signs of the Zodiac"? (a) Capricornus; (b) Pisces; (c) Aquarius; (d) Pegasus; (e) Sagittarius.

421. During what month of the year is the Sun primarily in the constellation Virgo (present epoch)?

422. What astronomical event occurs every year on or about the 22nd of September?

423. The Sun is at the vernal equinox on the following date (± 2 days): (a) 21 March; (b) 21 December; (c) 21 September; (d) 21 June.

424. Give approximate values for the right ascension and declination of the Sun on 21 December.

425. Criticize the following claim: All stars in a particular constellation (e.g., Orion) are at approximately the same distance from the Sun.

426. Explain why it is impossible to answer the following question: "What is the distance, in miles, between two stars five degrees apart in the sky?"

427. Cite several lines of evidence—physical rather than geometric—that the Earth is rotating.

428. As viewed from a northern mid-latitude station, what is the sense of rotation of circumpolar stars?

429. The period of rotation of the Earth is commonly said to be 24 hours. What, *precisely*, is the reference object with respect to which this period is measured?

430. Although the rotation of the Earth has been the classical basis for time-keeping it has been supplanted in modern precision work by (a) radioactive sources; (b) atomic clocks; (c) the revolution of the Galilean satellites of Jupiter; (d) lasers.

431. Write a one sentence definition of the sidereal day.

432. Define (a) the vernal equinox and (b) sidereal time.

433. The difference between the lengths of the solar and the sidereal day is principally due to (a) precession of the equinoxes; (b) perturbations of the Earth's orbit by the Moon; (c) gradual slowdown of the Earth's rotation; (d) orbital motion of the Earth about the Sun.

434. What is the equation of time?

435. A common unit of angular measure in scientific work is the radian. Define it and calculate the number of degrees in a radian.

436. An observer at ground level notes that the Washington Monument, whose height is 555 feet, subtends an angle of six degrees. How far is the base of the monument from the observer?

437. The principle of parallax is essential to the classical method of finding the (a) ratio of the Earth-Moon distance to the diameter of the Earth; (b) angular diameter of the Moon; (c) angle between two stars; (d) motion of the Moon with respect to the star field.

438. Two persons on the equator of the Earth and at nearly 180 degrees difference in longitude observe the Moon's position on the star field at the same moment. What is the difference in right ascension between the two apparent positions if the declination of the Moon is zero?

439. What is meant by the parallax of a star?

440. How can one distinguish a planet from a star by observation with the unaided eye?

441. Estimate the total number of stars that can be seen by the unaided eye on a clear, moonless night by counting those that you can see within a specified fraction of the sky.

442. How many stars can be seen by the unaided eye?

443. Given a photograph of an unspecified region of the sky, how does one proceed to identify the stars therein?

444. Why is it impossible to see stars during the daytime?

445. Fewer stars are visible on a night when the Moon is full than at other times of the month because of (a) moonlight scattered by the Earth's atmosphere; (b) increased cloudiness caused by the full Moon; (c) moonlight reflected from the ground; (d) refraction in the Earth's atmosphere.

446. The best astronomical seeing from the surface of the Earth permits a resolution of one second of arc. What is the minimum separation of two features on the surface of the Moon (in miles) such that they can be just distinguished from each other? (Mean distance to the Moon is 240,000 miles.)

447. Define a great circle on a sphere.

448. Define a small circle on a sphere.

449. List several spherical coordinate systems that are commonly used in astronomy.

450. The latitude of a place on the Earth is analogous to the _____ of a star on the celestial sphere.

451. The east longitude of a place on the Earth is analogous to the _____ of a star on the celestial sphere.

452. Using a labeled diagram, define North, South, East, and West at an arbitrary point on the star field.

453. What is the angle in degrees between two stars whose respective right ascensions and declinations are 22^h, 0 degree; and 20^h, 0 degree?

454. What is meant by the culmination of a star?

455. If an observer measures simultaneously the altitudes of three different stars having known right ascensions and declinations, the observer's position is (ideally) at the threefold intersection of the three corresponding small circles. What additional information is necessary to determine the latitude and longitude of the observer on the Earth? This problem is the classical one of celestial navigation.

456. All observers who measure the altitude of a particular star to be the same at a given moment are (a) on a meridian that passes through the sub-stellar point; (b) at the same longitude; (c) at the same latitude; (d) on a small circle centered on the sub-stellar point.

457. At the summer solstice all observers on a particular small circle on the Earth observe the Sun at the zenith at local noon. What is the name of this circle?

458. On 23 September, the Sun was at the autumnal equinox (beginning of autumn) (declination = 0 degrees, right ascension = 12^h). At what local apparent solar time on that date did a star rise if its declination was 0 degrees and its right ascension 06^h? Use the 24-hour time convention.

459. What is the angular length of a star trail near the celestial equator in a photograph with a fixed camera whose shutter is open for 12 minutes?

460. The altitude of Altair (declination = 9 degrees North) as it crosses a northern hemisphere observer's meridian is 41 degrees. What is the latitude?

461. At the winter solstice the altitude of the Sun at local noon in Iowa City (latitude 41.7 degrees North) is _____.

462. The altitude of Polaris is approximately equal to a northern hemisphere observer's (a) right ascension; (b) latitude; (c) east longitude; (d) west longitude.

463. If the Earth were a rigid (nondeformable) sphere and if it were caused to spin faster and faster, at what value of its rotational period would an object at the equator cease to rest on the Earth and become a synchronous satellite, in orbit just above the Earth's surface?

464. Where on the Earth are all stars above the horizon at one time or another?

465. A star is always on an observer's meridian at the same (a) local apparent solar time; (b) local mean solar time; (c) Greenwich Mean Time; (d) local sidereal time.

466. What are (a) the maximum and (b) the minimum altitudes of the Sun as observed at Thule, Greenland (latitude 78 degrees North) on 21 June?

467. Describe how to find "true" South (and hence "true" North) by observing a star in the southern sky.

468. What is the declination of a star that has an altitude from the south point of 25 degrees as it crosses the meridian of an observer at latitude 42 degrees North?

469. It is conceivable for a supersonic aircraft to fly in the equatorial plane of the Earth so that it is noon throughout the flight. In order to accomplish this the aircraft must (a) fly westward at 1,700; (b) fly eastward at 1,700; (c) fly eastward at 1,050; (d) fly westward at 1,300 km hr^{-1}.

470. The farthest south that an observer on the Earth can see Polaris is (a) the Arctic Circle; (b) the equator; (c) the Tropic of Cancer; (d) the Tropic of Capricorn.

471. What is a tropical year?

472. The calendar in contemporary use in most civilized nations is called the (a) Roman; (b) Julian; (c) Mayan; (d) Gregorian calendar.

473. Which one of the following years is a leap year? (a) 1973; (b) 1982; (c) 1986; (d) 1996.

474. One and only one of the following years is a leap year. Which one? (a) 1902; (b) 1986; (c) 1976; (d) 2100.

475. If the length of the tropical year were exactly 365^d 03^h, how often would we need a leap year? Once in ____ years.

476. During a 24-hour period the right ascension (measured in hours and minutes) of the Moon (a) changes by less than five minutes; (b) increases by about fifty minutes; (c) sometimes increases and sometimes decreases; (d) increases by about thirteen minutes.

477. On a particular day the Sun, the Moon, and a star (the latter two not being visible) cross an observer's meridian at the same time. On the following day the relative time order in which they cross the meridian (First, Second, Third) will be (a) Sun, Moon, star; (b) Moon, star, Sun; (c) star, Moon, Sun; (d) Moon, Sun, star; (e) star, Sun, Moon; (f) Sun, star, Moon.

478. What is the principal reason that we have seasons (i.e., summer, autumn, winter, and spring) on the Earth?

479. The principal cause of the run of the seasons on Earth (i.e., summer, autumn, winter, spring) is the (a) annual variation of the distance from the Sun to the Earth; (b) tilt of the rotational axis of the Sun to the normal to the ecliptic plane; (c) tilt of the rotational axis of the Earth to the normal to the ecliptic plane; (d) annual variation in the number and size of sunspots.

480. Days and nights are equal in duration throughout the year at the equator. Nonetheless a greater amount of solar energy is received there during one of the following months than during any one of the other three. Which one? (a) December; (b) May; (c) August; (d) September.

481. It is colder in the winter than in the summer for two principal reasons. The first is that the sunlit portion of the 24-hour day is shorter. The second is that (a) the Earth rotates more rapidly in the summer; (b) the Sun is farther away from the Earth in the winter; (c) the Sun's rays strike the Earth at smaller angles to its surface in the winter; (d) the cooler side of the Sun faces the Earth in the winter.

482. If the Earth's rotational axis were perpendicular to the plane of its orbit and if this orbit were a circular one of radius 1 AU (encircle one or more correct statements): (a) There would be no seasons. (b) The sidereal day would be equal to the solar day. (c) The equation of time would be zero. (d) The average temperature of the Earth would be about the same as it now is.

483. The length of the day would be equal to the length of the night throughout the year if, everything else being the same, (a) the rotational axis of the Sun were perpendicular to the ecliptic; (b) the Moon's orbit were coincident with the ecliptic; (c) the rotational axis of the Earth were perpendicular to the ecliptic; (d) the orbit of the Earth were circular rather than elliptical.

484. The pole of the ecliptic is in the constellation (a) Aquila; (b) Cassiopeia; (c) Draco; (d) Lyra.

485. At any instant there are two points on the Earth at which the local horizon coincides with the ecliptic. Where are they?

486. A northern hemisphere observer must be at a particular latitude in order that the ecliptic ever coincide with the horizon. What is this latitude?

487. The northern declination of the Moon is never greater than about ____ degrees.

488. As viewed from a latitude of 41 degrees North all circumpolar stars have (a) declinations less than 49 degrees North; (b) declinations greater than 49 degrees North; (c) right ascensions greater than 12^h; (d) declinations less than 41 degrees North.

489. The celestial equator crosses an observer's horizon at two points, one due East and one due West. At those points the celestial equator for a northern hemisphere observer at latitude φ is inclined to the horizontal at an angle (a) φ; (b) $90 - \varphi$; (c) $\varphi - 23.5$; (d) $\varphi + 23.5$ degrees.

490. On about what date does the winter night begin at the northern pole?

491. What is the approximate duration of the mean lunar day (analogous to the mean solar day) as measured in conventional units of hours and minutes?

492. If the Moon were used as the reference object for a terrestrial clock, the mean lunar day as measured in conventional units of time would be (a) 24^h 50^m; (b) 23^h 56^m; (c) 23^h 10^m; (d) 24^h 04^m.

493. Two points A and B are both at latitude 45 degrees North and are 700 miles apart. Find the difference in the longitudes of A and B.

494. A particular radio-broadcast satellite is in synchronous orbit in the equatorial plane at 6.6 Earth radii from the center of the Earth and at the longitude of an observer whose latitude is 42 degrees North. At what angle above the horizontal plane must the observer point the axis of a parabolic dish antenna for the best reception?

495. At the autumnal equinox the Sun sets due west at every latitude (except 90 degrees North or South). What are its azimuth and altitude at this moment?

496. An observer finds that the altitude of Polaris is 42 degrees. What is the observer's latitude?

497. A navigator's clock is in error by 04^m. What is the corresponding error in determination of longitude?

498. A properly designed sundial gives a direct reading of (a) local mean solar time; (b) local sidereal time; (c) local apparent solar time; (d) local apparent sidereal time.

499. Prove that the shadow-casting edge of the gnomon of a sundial must point at the celestial pole in order that its indication of local apparent solar time be independent of the Sun's declination.

500. Why does the rotational axis of the Earth precess in the star field?

501. The vernal equinox is now in Pisces whereas about 2,000 years ago it was in Aries. Briefly, what is the reason for this change?

502. The slow westward movement of the vernal equinox on the star field is attributable to (a) a gradual slowing of the rotation of the Earth; (b) continental drift; (c) gyroscopic precession of the axis of rotation of the Earth; (d) the cumulative effect of meteoric impacts on the Earth.

503. The rotational axis of the Earth precesses around a complete small circle on the celestial sphere in (a) 18.6; (b) 2,547; (c) 23.5; (d) 25,800 years.

504. The rotational axis of the Earth precesses uniformly on the celestial sphere along a small circle whose angular radius is 23.5 degrees. The length of time for one complete precession is 25,800 years. What will be the approximate declination of Polaris in the year 14,900 A.D.?

505. The period of precession of the Earth's axis is 25,800 years. About how many years does it take for the vernal equinox to move across one sign of the zodiac?

506. The present epoch has been called the "dawning of the Age of Aquarius". What is the astronomical significance of this phrase?

507. What is the brightest star in the sky, other than the Sun?

508. The constellation sketched here as seen in the northern sky is (a) Capricornus; (b) Cassiopeia; (c) Ursa Minor; (d) Pisces.

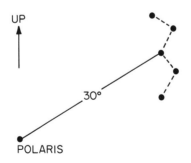

VIEW OF THE NORTHERN SKY –
EARLY EVENING, EARLY OCTOBER

509. Rigel is the name of a prominent star in the constellation (a) Orion; (b) Taurus; (c) Pegasus; (d) Cassiopeia.

510. In what constellation is each of the following bright stars:
     (a) Aldebaran _____
     (b) Rigel _____
     (c) Sirius _____
     (d) Capella _____

511. Aldebaran and the Pleiades are both in the constellation

     _____.

512. What is the brightest star in the constellation Pegasus?

513. The star Markab is in the constellation (a) Pegasus; (b) Aries; (c) Andromeda; (d) Cygnus.

514. Deneb is the brightest star in the constellation (a) Lyra; (b) Orion; (c) Pegasus; (d) Cygnus.

515. The brightest star in the constellation Cygnus is (a) Polaris; (b) Sirius; (c) Vega; (d) Deneb.

516. Name the constellations in which the following bright stars are found.
     (a) Polaris _____
     (b) Vega _____
     (c) Capella _____
     (d) Aldebaran _____

517. Altair is in the constellation (a) Canis Major; (b) Cygnus; (c) Orion; (d) Aquila.

518. Sketch the appearance of Ursa Major and the pole star (Polaris) in the northern sky as seen from your location in an evening in May.

519. Betelgeuse is listed in star catalogs as α Orionis. This designation means that it is (a) the northernmost star in the constellation Orion; (b) the brightest star in Orion; (c) the easternmost star in Orion; (d) the oldest star in Orion.

520. Cygnus is (a) a constellation of the Zodiac; (b) an asteroid; (c) a constellation that is never visible from mid-northern latitudes; (d) a constellation that contains the star Deneb.

521. Fill in the name of the constellation of which each of the following stars is a member:
(a) Markab   _____
(b) Vega   _____
(c) Altair   _____
(d) Arcturus   _____

522. What are the names of the two brightest stars in Orion?

523. What are the names of the two brightest stars in the constellation Gemini?

524. At about what time on the present date do the two pointer stars in Ursa Major point vertically downward toward Polaris? Use 24-hour clock convention and your local zone time.

525. The brightest star in the sky other than the Sun is (a) Canopus; (b) Vega; (c) Mars; (d) Sirius.

526. The Pleiades are in what constellation?

527. Spica is the brightest star in what constellation?

528. At local midnight on the present date which one of the following constellations is near the observer's meridian? (a) Lyra; (b) Pegasus; (c) Aquila; (d) Orion.

529. Sketch the current appearance of the Big Dipper (Ursa Major) in the northern sky at about 21^h, local zone time.

530. What would be your instructions to a friend who wishes to find Vega in the sky in mid-September?

531. Show by a simple diagram how to locate Polaris in the sky.

532. On the present date (a) what is the phase of the Moon? (b) at what time is Orion on the meridian? (c) what planets are accessible for observation and at what times of the night (most favorably)?

533. What is the name of the brightest star in the constellation Taurus?

534. In what constellation are the two bright stars Castor and Pollux?

535. Altair is the brightest star in the constellation (a) Pegasus; (b) Aries; (c) Aquarius; (d) Aquila.

536. Deneb is the brightest star in the constellation (a) Aquila; (b) Cygnus; (c) Lyra; (d) Ursa Major.

537. The cluster of stars called the Pleiades is in the constellation (a) Pegasus; (b) Orion; (c) Taurus; (d) Gemini.

538. The Pleiades are (is) (a) the brightest star in the southern hemisphere; (b) the fifth sign of the Zodiac; (c) a cluster of stars in the constellation Taurus; (d) a cluster of stars in the constellation Aries.

539. The bright stars Betelgeuse, Bellatrix, and Rigel are all in what constellation?

540. In what constellations are the following bright stars:
    (a) Capella   _____
    (b) Arcturus   _____
    (c) Rigel   _____
    (d) Altair   _____

541. Which one of the following is a prominent circumpolar constellation? (a) Aries; (b) Aquila; (c) Pisces; (d) Cassiopeia.

542. Give the names of two circumpolar constellations, as seen from mid-northerly latitudes.

543. In early evening in February, one of the following constellations is a prominent feature of the southern sky: (a) Ursa Major; (b) Cassiopeia; (c) Orion; (d) Pegasus.

544. Which three stars form the summer triangle?

545. At about what time on the present date do the two pointer stars in Ursa Major point vertically upward toward Polaris? Specify standard or daylight zone time.

546. The right ascension of Betelgeuse is 06^{h}. At what local solar time will it cross the meridian on 21 December? Use 24-hour convention.

547. The declination of Betelgeuse is 7.4 degrees North. What is the most southerly latitude on the Earth from which it can be seen?

548. For a specific observer, all stars having the same altitude at a given moment (a) have the same declination; (b) are at the same hour angle; (c) lie on a small circle on the celestial sphere centered on the observer's zenith; (d) have the same right ascension.

549. The second brightest star, other than the Sun, is Canopus. Its declination is 52.7 degrees South. In what range of latitudes can it be seen by an observer on the Earth? Note: Canopus, being only 14 degrees from the southern ecliptic pole, is an especially valuable reference star in the measurement and control of the orientation of interplanetary spacecraft.

550. The declination of Vega is 38 degrees 47 minutes North. At what latitude will it be directly overhead at some time during the 24-hour day?

551. A star whose declination is 10 degrees North crosses an observer's meridian at azimuth 0 degree and altitude 30 degrees. What is the observer's latitude?

552. On 22 June a star is observed to be in the direction of the vernal equinox; its speed is found by means of the Doppler effect to be 35 km s^{-1} toward the Earth. Six months later on 22 December the same star is found to have a speed of 25 km s^{-1} away from the Earth. What is the speed of the star with respect to the Sun?

553. The full Moon crosses the meridian of a northern hemisphere observer at a much greater altitude in the winter than in the summer. Explain.

554. The altitude of Altair (declination = 9 degrees North) as it crosses a southern hemisphere observer's meridian is 41 degrees. What is the observer's latitude?

555. The zone time in the 90 degree West zone is 08^h CST on a Saturday. What is the central longitude of the time zone in which the standard zone time at the same moment is 01^h on Sunday?

556. In early December the star Capella (a) crosses the meridian in the early evening; (b) rises in the southeast at about midnight; (c) is low in the western sky during evening twilight; (d) is high in the northeastern sky in the early evening.

557. The west longitude of Green River, Wyoming is 109.5 degrees and of Iowa City, 91.5 degrees. What is the local mean solar time in Iowa City when it is 13^h 20^m in Green River?

558. Vega (right ascension 19^h, declination 39 degrees North) is observed on 21 June to cross the local meridian at an altitude of 75 degrees above the South point at 23^h Greenwich Mean Time. Find the latitude (specify N or S) and longitude (specify E or W) of the point of observation. Neglect the equation of time.

559. The approximate longitudes of New Delhi, India, and Iowa City are 75 degrees East and 90 degrees West, respectively. When it is 09^h CST on Sunday in Iowa City what are the time (24-hour convention) and day of the week in New Delhi (using standard zone times)?

560. When the standard zone time 90 degrees West longitude is 18^h on Thursday, what are the standard zone time (45 degrees East longitude) and day of the week in Baghdad, Iraq?

561. A particular star rises at 20^h Local Mean Time on 25 October. At about what time will it rise on 25 November? (a) 18^h; (b) 21^h; (c) 19^h; (d) 22^h.

562. Sirius (right ascension 07^h, declination 17 degrees South) is observed on 22 March to cross the local meridian at an altitude of 40 degrees above the South point at 01^h Greenwich Mean Time. Find the longitude (and label it E or W) and latitude (and label it N or S) of the point of observation.

563. The declination of Canopus, the second brightest star, is 53 degrees South. The most northerly latitude from which it can be seen under ideal conditions at sometime during the year is (a) 37; (b) 53; (c) 47; (d) 23.5 degrees.

564. The declination of Altair is 9 degrees North. What is its altitude when it crosses the meridian of an observer at 56 degrees North latitude?

565. At local noon on 21 March you are temporarily blinded by the image of the Sun reflected from the windshield of a car approaching you from the north on a level highway. The windshield is a flat piece of glass inclined at 20 degrees from the vertical. What is your latitude?

566. The right ascension of Aldebaran is 04^h 36^m. What is the local sidereal time when its hour angle is 02^h 36^m?

567. When the vernal equinox is on the meridian at longitude 90 degrees West, the Greenwich sidereal time is (a) 18^h; (b) 09^h; (c) 06^h; (d) 05^h.

568. The latitude of Thule, Greenland is 78 degrees North. Between approximately what dates of the year does the Sun never appear above the horizon?

569. The latitude of Thule, Greenland is 78 degrees North. On about what date following the Arctic summer does the Sun first set in the northern sky?

570. On 22 September an observer noted that the Sun rose at 12^h and set at midnight Greenwich Mean Time. Hence, the longitude was about (a) 90 degrees West; (b) 90 degrees East; (c) 45 degrees West; (d) 180 degrees West.

571. When it is noon in London, what time is it in Chicago? Use standard zone times.

572. A star that rises at an azimuth of 75 degrees will set at an azimuth of (a) 285; (b) 75; (c) 265; (d) 255 degrees.

573. A fixed camera is pointed upwards toward the northern celestial pole and the shutter is held open for, say, 2 hours. The resulting photograph shows star trails as arcs of circles centered on the pole (near Polaris). Describe a simple technique for (a) labeling the photograph as to sense of rotation and (b) determining the sidereal period of rotation of the Earth.

574. The latitudes of Iowa City and Des Moines are about the same but their respective longitudes are 91 degrees 32 minutes West and 93 degrees 27 minutes West. On a day on which the Sun rises at 06^h 35^m CST in Iowa City, at what time does it rise in Des Moines?

575. The right ascension of Spica is $13^h 23^m$. During what month of the year will it be on an observer's meridian at local midnight?

576. At the winter solstice (21 December) the declination of the Sun is 23.5 degrees South. What is the Sun's altitude as it crosses the meridian of an observer at 50 degrees North latitude?

577. What is the hour angle of the mean Sun in New York (longitude = 75 degrees West) when it is 18^h Greenwich Mean Time?

578. A star that rises at an azimuth (clockwise from north) of 50 degrees will set at an azimuth of (a) 310; (b) 230; (c) 130; (d) 140 degrees.

579. An observer at latitude 66.5 degrees North notes that the Sun is above the horizon for only about 0.5 hour each day. The month of the year is one of the following: (a) March; (b) June; (c) December; (d) July.

580. Which one of the following statements is *not* true at the time of the autumnal equinox? (a) The right ascension of the Sun is 12^h. (b) The declination of the Sun is 0 degrees. (c) The length of the day is equal to the length of the night at every point on the Earth's surface (exclusive of the poles). (d) The Sun is in Aries.

581. At what Greenwich Mean Time is it the same day of the week throughout the Earth?

582. What is the astronomical significance of the Arctic Circle (latitude 66.5 degrees North)?

583. Some opponents of Daylight Saving Time consider Standard Time to be "God's time". Discuss this point of view from an astronomical standpoint.

584. At one moment during the year the local sidereal time (hour angle of the vernal equinox) is identical to the local apparent solar time. What is the astronomical name for that moment?

585. At what time of the year is the local sidereal time equal to the local solar time? (a) summer solstice; (b) vernal equinox; (c) autumnal equinox; (d) winter solstice.

586. The right ascension of Rigel is 05^h 12^m. On about what date of the year will it be on an observer's meridian at local midnight?

587. What is the greatest southerly declination of a star that can be seen at some time during the year from a station at latitude 30 degrees North?

588. The latitude of a northern hemisphere observer is approximately equal to (a) the altitude h of Polaris; (b) h + 90; (c) 90 − h; (d) 180 − h degrees.

589. What is the declination of a star that has an altitude from the south point of 35 degrees as it crosses the meridian of an observer at latitude 35 degrees North?

590. The longitude of Iowa City is 91.5 degrees West. What is the local mean solar time there when Central Daylight Time is 09^h 40^m?

591. Give your best estimates of the altitude and azimuth of the Moon last night at 22^h.

592. At the autumnal equinox a star having a right ascension of 06^h and a declination of 0 degrees rises at a local solar time of (a) 18^h; (b) 06^h; (c) 12^h; (d) 24^h.

593. When the Greenwich Mean Time on 21 September is 01^h, the sidereal time at Greenwich is approximately (a) 01^h; (b) 23^h; (c) 13^h; (d) 11^h; (e) none of the above.

594. At full Moon, the difference in right ascensions of the Sun and the Moon is (a) 24^h; (b) 00^h; (c) 09^h; (d) 12^h.

595. A particular star rises at 21^h 30^m on 25 June. On 25 July it will rise at (a) 20^h 26^m; (b) 23^h 34^m; (c) 19^h 26^m; (d) 16^h 20^m.

596. A star that rises at an azimuth (clockwise from north) of 120 degrees will set at an azimuth of (a) 210; (b) 60; (c) 120; (d) 240 degrees.

597. On 21 March the Sun is at the vernal equinox (beginning of spring) (right ascension $= 00^h$, declination $= 0$ degrees). At what local apparent solar time on that date will a star rise if its declination is 0 degrees and its right ascension is 12^h? Use 24-hour clock convention.

598. On 10 October, Orion is on an observer's meridian at 05^h 30^m Central Daylight Time. At about what Central Standard Time will it be on the meridian on 10 December?

599. As viewed by an observer at longitude 91.5 degrees West the hour angle of a given star is 02^h 30^m when the Greenwich sidereal time is 22^h. Therefore, the right ascension of the star is (a) 13^h 24^m; (b) 08^h 36^m; (c) 10^h 36^m; (d) 18^h 24^m.

600. What is the approximate local apparent solar time when the first quarter Moon is on an observer's meridian? Use 24-hour convention.

601. What is the age (number of days since new Moon) of the Moon if it crosses an observer's meridian at 21^h local apparent solar time?

602. The International Date Line is approximately coincident with (a) the meridian of the Fiji Islands; (b) the Tropic of Capricorn; (c) the Tropic of Cancer; (d) the meridian of Greenwich, England.

603. An observer notes that the Sun rose at 11^h GMT and set at 23^h GMT. Hence, the observer's longitude is about (a) 165 degrees West; (b) 75 degrees West; (c) 75 degrees East; (d) 60 degrees West. Neglect the equation of time.

604. A star that rises at 23^h 20^m CDT on 1 August will rise at _____ CDT on 1 September.

605. If an observer's latitude is φ degrees North, the altitude of the Sun at local noon at the vernal equinox is (a) 2φ; (b) φ; (c) $90 - \varphi$; (d) $\varphi - 90$ degrees.

606. On a particular date at a particular location, the Sun sets at an azimuth of 238 degrees. At what azimuth does it rise?

607. The Greenwich hour angle of the mean Sun is 02^h and the Greenwich day is Thursday. Hence, at Iowa City the Central Standard Time and day are (a) 08^h Thursday; (b) 20^h Wednesday; (c) 20^h Thursday; (d) 18^h Wednesday.

608. A star on the celestial equator has an hour angle of 18^h at a particular moment. Hence, its azimuth is (a) 180; (b) 90; (c) 0; (d) 270 degrees.

609. Approximate the longitudes of Ho Chi Minh City and Iowa City as 105 degrees East and 90 degrees West, respectively. When it is 09^h local mean solar time on a Wednesday in Iowa City what are the local mean solar time and day of the week in Ho Chi Minh City? Use 24-hour time convention.

610. If the Sun is on an observer's meridian at 10^h Greenwich Mean Time, what is the observer's longitude? Neglect the equation of time.

611. What is the local mean solar time in Omaha (longitude 96 degrees West) when the Central Standard Time is 11^h 00^m?

612. The shortest day of the year occurs on 22 December, the winter solstice, but at mid-northern latitudes the earliest sunset occurs on about 9 December and the latest sunrise, on about 4 January, both measured in local mean solar time. Explain.

613. An observer notes that the Sun rises at 05^h and sets at 17^h Greenwich Mean Time. Hence the observer's longitude is (a) 15 degrees East; (b) 75 degrees East; (c) 105 degrees West; (d) 15 degrees West. Neglect the equation of time.

614. In tables for navigators (e.g., *The Air Almanac*, a semiannual publication of the U.S. Naval Observatory), values of the Greenwich Hour Angle (GHA) of the Sun and other prominent celestial objects and of the Greenwich Hour Angle of the vernal equinox (GHA γ) are tabulated as a function of Greenwich Mean Time (GMT). Write down the relationship among these quantities and the right ascension (α) of the object.

615. When Greenwich Mean Time is 16^h on Monday, what is the mean solar time and day of the week at 135 degrees East longitude?

616. A particular star has a right ascension of $13^h 52^m$. What is its hour angle for an observer whose local sidereal time is $10^h 25^m$?

617. When it is 12^h CST on 22 January in Chicago (90 degrees West meridian time), the standard zone time and date in Japan (135 degrees East meridian time) are (a) 21^h on 22 January; (b) 21^h on 23 January; (c) 03^h on 23 January; (d) 03^h on 21 January; (e) 21^h on 21 January.

618. Suppose that you are on a ship at sea and maintain your chronometer on Greenwich Mean Time. You observe the Sun rising at 08^h and setting at 21^h. Your longitude is (a) 37.5 degrees West; (b) 37.5 degrees East; (c) 2.5 degrees West; (d) 45 degrees West. Assume that the equation of time is zero.

619. How is it possible to guide a ship across the Atlantic Ocean from Europe to North America without any knowledge of time?

620. What is the essential information that a navigator needs in order to determine position (longitude and latitude) from celestial observations?

621. In the determination of longitude and latitude by observation of the altitudes of known stars: (a) It is essential to also know Greenwich Mean Time (GMT). (b) It is helpful but not essential to know GMT. (c) Knowledge of GMT is unnecessary if the altitudes of three or more stars are observed simultaneously. (d) Knowledge of GMT is unnecessary if a particular star is observed repeatedly at known intervals of time.

622. Two points A and B on the surface of a sphere can be connected by an infinite number of small circles but by one and only one great circle. Prove that the shortest distance between A and B is along the great circle.

Notes:

(1) The author's answers to various problems in this chapter and other chapters adopt $24^h 50^m$ as the average time interval between successive meridian transits of the Moon; and $23^h 56^m$ as the length of the sidereal day.

(2) In visualizing problems involving time, it is convenient to adopt the geocentric point of view with all objects on the celestial sphere revolving clockwise (westerly) around the nonrotating Earth. A circular time dial with tick marks at intervals of 15 degrees (1 hour) and with the northern celestial pole at the center is recommended as a helpful device. On such a dial hour angles increase in the clockwise (westerly) sense and right ascensions increase in the counterclockwise (easterly) sense.

CHAPTER 5

Other Planets, Their Satellites, and Rings

623. Which one of the nine major planets is missing from the following list: Mercury, Uranus, Pluto, Earth, Saturn, Mars, Jupiter, Venus?

624. The brightness of a planet of a given size and albedo as seen from the Sun varies (a) directly as the square; (b) inversely as the square; (c) directly as the three-halves power; (d) inversely as the fourth power of its distance.

625. The radiative equilibrium temperature of an inert object in space is directly proportional to the fourth root of the intensity of light that illuminates it. Starlight is less intense than sunlight at 1 AU by a factor of 1×10^8. Calculate the temperature of an object in interstellar space, remote from any star, by comparison with the temperature of the Earth.

626. The nine major planets of the solar system are often classified into two types: terrestrial and Jovian. What is the basis for this classification and which planets are which? Into which classification would the minor planets fall?

627. Which one of the following is a "terrestrial" planet? (a) Jupiter; (b) Uranus; (c) Neptune; (d) Mars.

628. Name the five planets (other than the Earth) that can be seen by the unaided eye at some time during a year.

629. Of what planet or planets are *all* of the following statements true?
    (1) It moves along its orbit in a counterclockwise sense as viewed from the northern celestial pole.
    (2) Its rotational period is less than thirty hours.
    (3) Its mass is greater than that of the Earth.
    (4) It has more than one natural satellite.

630. *All* of the following four statements are true for only *one* of the nine major planets. (a) Its axis of rotation is inclined at less than 30 degrees to the normal to its orbital plane. (b) Its atmosphere is principally carbon dioxide. (c) Its mean density is between 5 and 6 g cm^{-3}. (d) Its surface is not visible because of heavy cloud cover. Which one is it?

631. All major planets are of approximately spherical shape. The principal physical reason for this is (a) gravitational attraction of every element of the planet by every other element; (b) erosion by meteoric impacts; (c) the average gravitational effect of satellites of the planet; (d) gravitational attraction of every element of the planet by the Sun.

632. Discuss the fact that all large astronomical objects (e.g., the Sun, stars, and major planets) are approximately spherical whereas minor planets (asteroids), nuclei of comets, and small satellites of planets are of irregular shapes. Estimate the maximum radius of a rocky object that can have a markedly non-spherical shape.

633. Estimate the greatest physically possible height of a mountain on a solid-body planet. Compare your estimate with actual data on the Earth, Mars, and Venus (and the Moon).

634. Suppose that a tunnel is bored through the Earth in a straight line from New York to San Francisco and equipped so that a car can coast through it without friction or air resistance. Discuss the motion of the car after its release from rest at the mouth of the tunnel in New York.

635. Why is the deep interior material of a planet under high pressure?

636. The gross chemical constituency of a planet, an asteroid, or a planetary satellite is revealed by a single characteristic quantity. What is that quantity?

637. The mean density of a planetary body provides the most important single item of information on (a) its age; (b) its rate of rotation; (c) its internal composition; (d) its orbital speed.

638. The distinction between the terrestrial planets and the Jovian planets is based primarily on their (a) synodic periods of revolution; (b) mean densities; (c) rotational periods; (d) diameters.

639. *All but one* of the following is a terrestrial (i.e., Earth-like) planet: (a) Mars; (b) Mercury; (c) Venus; (d) Saturn.

640. Name the five known outer planets of the solar system.

641. Which planet has the most prominent system of rings?

642. An astronomer reports the discovery of a planet of the brightness of Uranus moving northward through the constellation Ursa Major at the rate of five degrees per year. A tentative professional assessment of this report would include *all but one* of the following points. Which one? (a) The corresponding orbit has a semimajor axis of about 17 AU. (b) The sidereal period of the planet is about 72 years. (c) The reported data are incompatible with Kepler's and Newton's laws. (d) The implied orbit is markedly different than that of any previously known planet.

643. Suppose that a newly discovered planet is observed to move eastward on the star field near the ecliptic at an average rate of nine degrees per year. Hence its orbit (assumed to be circular) lies (a) between the orbits of Mars and Jupiter; (b) beyond the orbit of Pluto; (c) between the orbits of Saturn and Uranus; (d) between the orbits of Jupiter and Saturn.

644. During its orbital motion, the Moon rotates so that it keeps the same face toward the Earth. All close satellites of Mars, Jupiter, and Saturn exhibit similar behavior. What are the physical causes of this behavior?

645. Within the solar system, the surface temperature of a body varies inversely as the square root of its distance from the Sun, being about 270 degrees Kelvin at 1 AU. Which one of the following substances (with their freezing temperatures in degrees Kelvin) would, if present, be in a gaseous state in Pluto's atmosphere at 40 AU? (a) CH_4, 91; (b) H_2O, 273; (c) H_2, 14; (d) N_2, 63.

646. What physical conditions determine whether or not a planet or a satellite can retain an atmosphere?

647. The two principal factors that determine the capability of a planet or a satellite for retaining an atmosphere are its temperature and (a) its rotational period; (b) the roughness of its surface; (c) the gravitational escape speed from its surface; (d) its orbital speed.

648. One of the two principal molecular constituents of the Earth's atmosphere is essential to human life but is almost totally absent in the atmospheres of Venus, Mars, Jupiter, and Saturn. What is it?

649. The density of the Earth's atmosphere is less by a factor of about two for each increase of altitude by 5.5 km. Hence the atmospheric density at an altitude of 33 km (108,000 ft) is less than that at sea level by the factor _____.

650. Name four chemical *elements* that are the dominant constituents of the atmospheres of all planets that have substantial atmospheres.

651. Four hypothetical planets have the following surface values of temperature T and speed of escape v.

	T	v
Alpha	300 degrees Kelvin	10 km s^{-1}
Beta	200	12
Gamma	500	2
Delta	100	20

Which one is least likely to have an atmosphere?

652. The melting (or sublimating) temperatures of some common substances in degrees Kelvin are as follows:

O_2	54	NH_3	195
H_2O	273	CH_4	91
CO_2	195	N_2	63
H_2	14		

Which of these substances would be most likely to exist in solid form on the surface of a satellite of Saturn?

653. Two satellites of Jupiter (Io and Ganymede) and one of Saturn (Titan) are now known to have measurable atmospheres, whereas the Moon, which is of similar size, does not. What is the single most plausible reason for this difference?

654. Only one of the following planets has a strong magnetic field. Which one? (a) Jupiter; (b) Mars; (c) Venus; (d) Mercury.

655. Only the Earth and four other planets are known to have radiation belts. Name the other four.

656. Name five planets that have intense radiation belts and extensive magnetospheres.

657. The Roche limit is concerned with (a) the existence of Lagrange points in the Sun-Jupiter system; (b) the breakup of objects in close orbits about a planet or the Sun; (c) the greatest possible speed that a comet can have; (d) the greatest possible rate of rotation of a planet before its disintegration.

658. The Roche limit for the minimum radius of the orbit of a large satellite of a planet of radius a is about 2.5 a. The basic physical reason for the existence of such a limit involves consideration of (a) Kepler's first law; (b) tidal forces; (c) Kepler's second law; (d) the period of rotation of the satellite.

659. The Roche limit for the minimum radius of the orbit of a satellite of a planet of radius a is about 2.5 a. The basic physical reason for the existence of such a limit involves consideration of (a) tidal forces; (b) Bode's law; (c) rotational period of the planet; (d) Kepler's third law.

660. Give an elementary derivation of the Roche limit treating the orbiting satellite as consisting of two solid spheres held together by their mutual gravitational attraction and pulled apart by tidal forces.

661. Sketch the essentials of how the masses of the Earth and the other eight planets have been determined.

662. How can the mass of a planet that has no natural satellites (e.g., Venus) be determined?

663. In the pre-spacecraft epoch, the mass of Mercury was found by (a) analysis of the gravitational perturbation that it produces on the orbital motions of Venus and other planets; (b) the radar determination of its sidereal period of rotation; (c) measurement of the eccentricity of its orbit; (d) observation of the sidereal periods of revolution of its two small satellites.

664. A nonrotating object is moving along a straight line at a constant speed v of 1 km s^{-1} past an observer who is located at 1 km from this line. Calculate as a function of time (a) the radial component of the object's velocity relative to the observer and (b) the apparent rate of rotation of the object as viewed by the observer.

665. Describe the radar technique that Pettingill and Dyce used to determine the rotational period of Mercury.

666. Describe how one can determine the radial component of the velocity of Mercury with respect to the Earth and its period of rotation by using the Doppler effect in reflected radar pulses.

667. The planet Mercury rotates in the prograde sense with a sidereal period of 58.65 days and revolves about the Sun with a sidereal period of 87.969 days. Assuming a circular orbit what is the length of time between successive meridian transits of the Sun for an observer on Mercury?

668. The theoretical surface temperature of an inert body in the solar system varies as the reciprocal square root of its distance from the Sun. If the planet Mercury with a mean surface temperature of 450 degrees Kelvin were moved from its present orbit (a = 0.39 AU) to the orbit of Neptune (a = 30 AU), what would be its mean surface temperature?

669. Show by a simple drawing the disc of Venus when it is at maximum elongation East. Draw a circle, with north up and west to the right, and blacken the portion of the planetary disc that is not illuminated by the Sun.

670. The sidereal and synodic periods of revolution of Venus are 225 and 584 days, respectively. The sidereal period of its retrograde rotation is 243 days and hence the length of the Venusian "day" is 117 Earth days. Prove that the same side of the planet faces the Earth at every inferior conjunction.

671. Venus rotates in the retrograde sense (i.e., clockwise as viewed from the northern ecliptic pole) with a sidereal period of 243 days and with its axis of rotation approximately perpendicular to its orbital plane. The sidereal period of its prograde revolution about the Sun is 225 days. An observer on Venus finds the Sun to be on the meridian at t = 0. How many days will elapse before the next time that the Sun is on the meridian? A graphical analysis is adequate to make a choice among the following values: (a) 468; (b) 247; (c) 117; (d) 18 days.

672. The surface temperature of Venus (about 720 degrees Kelvin) has been determined by (a) microwave radiometers; (b) infrared radiometers; (c) charged particle detectors; (d) calculation from its observed optical albedo.

673. Two important discoveries in planetary astronomy are that the surface temperature of Venus is about 720 degrees Kelvin and that atmospheric pressure at its surface is about (a) 90; (b) 3; (c) 6; (d) 10 times as great as that at the surface of the Earth.

674. The sidereal period of rotation of Venus has been found to be 243 days (retrograde) despite the fact that the dense cloud cover of the planet prevents the observation of its solid surface by optical telescopes. How has this period been determined?

675. A hypothetical planet has a sidereal period of revolution of 240 days and rotates in the retrograde sense with a sidereal period of 180 days. Find the length of time between successive "noons" at a given point on the planet's surface.

676. The original discovery of high mountains on the surface of Venus was accomplished by (a) infrared photography by Mariner 5 during a close flyby in 1967; (b) Earth-based radar telescopes; (c) ultraviolet photography from the Orbiting Astronomical Observatory; (d) rare glimpses by Earth-based optical telescopes through gaps in the cloud cover.

677. The planet Mars is said to rotate once in 24^h 37^m. How has this period been determined?

678. As seen from the Earth, the brightness of Mars at opposition is about (a) 4.8; (b) 23; (c) 1.52; (d) 2.3 times greater than at conjunction.

679. The most abundant molecular constituent of the Martian atmosphere is (a) CO_2; (b) O_2; (c) N_2; (d) H_2O.

680. The bright polar caps of Mars have been found to consist of (a) a thin covering of vegetation; (b) dust clouds; (c) a thin covering of frozen carbon dioxide and water; (d) smoke from active volcanoes.

681. One of the findings by the Mariner 9 spacecraft in orbit around Mars was that (a) the atmosphere is mainly nitrogen; (b) there are large swampy areas; (c) active volcanoes are erupting large quantities of fine dust; (d) most craters appear to be of meteoric origin.

682. The spectrum of sunlight reflected from Mars reveals information about *all but one* of the following matters: (a) composition of its solid surface; (b) gaseous composition of its atmosphere; (c) temperature of its interior; (d) variability of water vapor in its atmosphere.

683. Mars' satellite Phobos revolves about the planet with a sidereal period of about 7.7 hours. The planet rotates in the same sense with a period of about 25 hours. What is the interval of time between successive transits of Phobos across a given Martian meridian?

684. Based on present scientific knowledge, the most favorable site in the solar system for the development of biological material is thought to be (a) the surface of Venus; (b) the surface of the Moon; (c) the interior of Jupiter; (d) the surface of Mars.

685. Instruments on the two Viking Landers on Mars (a) discovered micro-organisms in the surface soil; (b) found oxygen to be the principal constituent of the atmosphere; (c) established the absence of complex organic molecules in surface soil; (d) were disabled by meteoric bombardment.

686. Briefly, what are your thoughts on the likelihood of the existence of living organisms on one of the other planets or its satellites?

687. A popular book called *The Jupiter Effect* attributes exceptionally high tides on the Earth and an enhanced probability of earthquakes to times at which Jupiter is in opposition. Discuss the plausibility of such claims.

688. The mean density of Jupiter is 1.33 g cm^{-3}. What is the principal implication of this one fact?

689. Among the nine major planets Jupiter is distinguished by one of the following characteristics: (a) It is the only one to rotate in the retrograde sense. (b) It is the largest. (c) It has no natural satellites. (d) It is the most distant from the Sun.

690. The mass of Jupiter can be determined by observation of any one of its satellites. Which one of the following quantities is essential to such a determination: (a) period of revolution of the satellite; (b) mass of the satellite; (c) eccentricity of its orbit; (d) inclination of its orbit?

691. The mass of Jupiter can be found by (a) comparing its sidereal and synodic periods of revolution; (b) comparing its sidereal period of revolution with that of the Earth; (c) measuring its period of rotation and oblateness; (d) none of the above methods.

692. The classification of Jupiter as a strongly magnetized planet is based on one of the following types of observational evidence: (a) in situ magnetic field measurements near the planet; (b) its shape; (c) spectral features in sunlight reflected from its surface; (d) non-Keplerian motion of its satellites.

693. The radii of Jupiter and the Earth are 71,400 and 6,372 km, respectively. About how many earths can be enclosed by the same volume of space as that occupied by Jupiter? (a) 1,400; (b) 11; (c) 126; (d) 394.

694. The mass of Jupiter is 318 times that of the Earth. What is the radius of the circular orbit of a satellite of Jupiter that would have a sidereal period of revolution of 27 days? (a) 6.8×10^6; (b) 1.2×10^8; (c) 3.8×10^7; (d) 2.6×10^6 km. The radius of the Moon's orbit is 384,400 km.

695. Using the approximations that Jupiter's visible radius is eleven times as great as that of the Earth and that its mean density is one-fourth as great, estimate the acceleration due to gravity at its cloud tops.

696. The visible radius of Jupiter is eleven times that of the Earth and its mass is 318 times as great. Find the ratio of the acceleration due to gravity at the cloud tops of Jupiter to that at the surface of the Earth.

697. Studies of Jupiter show that (a) its surface is solid material; (b) there is evidence for widespread biological activity in its atmosphere; (c) the formation of organic materials in its atmosphere is chemically impossible; (d) its atmosphere contains complex turbulent cloud systems.

698. Concerning Jupiter, which one of the following statements is *not* true? (a) Jupiter is a strongly magnetized planet. (b) Jupiter has an internal heat source of the same nature as that mainly responsible for the Sun's heat. (c) The many satellites of Jupiter are analogous in the Keplerian sense to the system of planets in orbit about the Sun. (d) Hydrogen and helium are the most abundant elements in Jupiter.

699. Give a brief summary of the satellites of Jupiter.

700. There are many active volcanoes on the Earth, but only one other object in the solar system is known to have active volcanoes. What is that object?

701. The rapid rotation of Jupiter in the prograde sense (i.e., counterclockwise as viewed from the northern ecliptic pole) (a) tends to speed up its orbital motion; (b) tends to slow down its orbital motion; (c) has no effect on its orbital motion; (d) improves its likelihood of capturing an asteroid.

702. It has been discovered recently that the two outer Galilean satellites of Jupiter—Ganymede and Callisto—have mean densities of about 1.9 g cm^{-3}. Hence they probably consist principally of (a) rocky material similar to that in the Moon; (b) iron, nickel, cobalt, and other heavy elements; (c) ices of ammonia, carbon dioxide, and water with an admixture of heavier elements; (d) hydrogen and helium.

703. The largest satellite of a planet in the solar system is Ganymede. Of what planet is it a satellite?

704. The four outermost satellites of Jupiter are distinguished from the twelve others by one of the following characteristics: (a) They have atmospheres of helium. (b) They revolve in retrograde orbits. (c) They are much larger. (d) Their orbits are much less eccentric.

705. Which one of the following is *not* a satellite of Jupiter? (a) Titan; (b) Europa; (c) Ganymede; (d) Callisto.

706. Suppose that the circular orbit of a small satellite of Jupiter has a radius of 700,000 km and is inclined at an angle of twenty degrees to the orbital plane of Jupiter. What is the approximate interval of time between successive eclipse seasons?

707. The first spacecraft to make a close encounter with Saturn was (a) Helios 1; (b) Voyager 1; (c) Pioneer 11; (d) Voyager 2.

708. Which one of the following is a satellite of Saturn? (a) Callisto; (b) Titan; (c) Icarus; (d) Hidalgo.

709. The magnetic field and magnetosphere of Saturn were discovered and investigated in 1979 by scientific instruments on (a) Mariner 9; (b) Pioneer 11; (c) Voyager 1; (d) Ulysses.

710. Which one of the following lines of evidence (before the epoch of spacecraft) established that the rings of Saturn are not thin rigid discs (perhaps containing holes) but consist of individual particles in essentially independent Keplerian orbits about the planet: (a) observation of the radial dependence of the Doppler shift of absorption lines in reflected sunlight; (b) determination of the optical brightness of the rings at various phase angles; (c) determination of the deficiency of blue light in sunlight reflected from the rings; (d) observation of stars through the rings.

711. The rings of Saturn (a) consist of large and small pieces of ice; (b) are at least a few thousand kilometers thick; (c) consist of thin solid sheets of a whitish metal; (d) are edge-on to the Earth at every opposition.

712. A large number of small objects are initially in prograde orbits of miscellaneous eccentricities and inclinations about an oblate planet (e.g., Saturn). Prove that all of these objects and their fragmentation products will end up eventually in circular orbits in the equatorial plane of the planet (e.g., Saturn's rings).

713. A discovery of Voyager 1 was that (a) all close satellites of Saturn revolve in retrograde orbits; (b) Titan rotates with a sidereal period of about 20 hours; (c) Titan has a dense atmosphere, principally of nitrogen; (d) Saturn has a solid surface, partially obscured by a thin haze.

714. The existence of thin rings of particulate matter but no large satellites within a certain radial distance from Saturn is generally considered to confirm a theoretical calculation on tidal forces originally done by (a) Roche; (b) Schiaparelli; (c) Dollfus; (d) Einstein.

715. In a comparative summary of the orbits and other properties of the planets, the most unusual characteristic of Uranus is (a) the eccentricity of its orbit; (b) its high rotational rate; (c) the orientation of its rotational axis; (d) its low albedo.

716. The rotational axis of Uranus is inclined at an angle of eight degrees to the plane of its orbit and its period of revolution about the Sun is 84 years. Describe the seasons on Uranus.

717. The planet Uranus rotates about an axis lying nearly in the plane of its orbit about the Sun. It makes one revolution about the Sun in 84 years. Hence, the Sun crosses Uranus' celestial equator once every (a) 2; (b) 84; (c) 42; (d) 21 years.

718. The rotational axis of Uranus lies approximately in the plane of its orbit, and its period of revolution about the Sun is 84 years. Hence the interval of time between vernal and autumnal equinoxes on Uranus is (a) 168; (b) 84; (c) 42; (d) 21 years.

719. In March 1977 it was discovered that Uranus has five thin rings (analogous to those of Saturn). Briefly, what observational technique was used?

720. Tombaugh's careful survey of the sky would have revealed a planet as large as Neptune at a distance of 240 AU from the Sun. To about how great a distance could he have discovered a planet of the same size as the Earth? The radius of Neptune is about four times that of the Earth.

721. Pluto was first observed by (a) Kuiper; (b) Schiaparelli; (c) Tombaugh; (d) Lowell.

722. If one examines estimates of the radius of Pluto over the period since its discovery by Tombaugh in 1930, one finds that they have diminished by a large factor. Explain.

23. The planet Pluto is about forty times as far from the Sun as is the Earth. Therefore, the ratio of the intensity of sunlight at Pluto to that at the Earth is (a) 1/40; (b) 1/6; (c) 1/1,600; (d) 1/2.

24. The diameter of Pluto has been estimated by (a) observing its period of rotation; (b) measuring its brightness; (c) observing the Doppler shift of lines in the spectrum of sunlight that it reflects; (d) observing its gravitational influence on the orbit of Mercury.

25. The rotational period of Pluto has been found by (a) measuring the ratio of its polar diameter to its equatorial diameter; (b) observing the Doppler shift in a reflected radar signal; (c) photographing the movement of a prominent crater across its visible disk; (d) measuring the cyclic variation of its brightness.

26. What is the name of Pluto's only known satellite?

27. By what observational technique has the period of rotation of Pluto been measured?

28. The presently accepted value of the rotational period of Pluto is 6.39 days. How has this period been determined?

29. Arrange the following six solar system bodies in order of size, largest first—smallest last: Pluto, Triton, Titan, Ganymede, Mercury, Moon.

CHAPTER 6

Asteroids, Comets, and Meteoroids

730. An asteroid is (a) a small star; (b) a minor planet; (c) a large gas cloud; (d) a small satellite of a planet.

731. What is an asteroid? Where are most of the asteroids?

732. The usual distinction between a planet and an asteroid is based on (a) orbital period; (b) mean density; (c) size; (d) date of discovery.

733. About how many asteroids have been identified and cataloged?

734. Describe the orbit of an average asteroid.

735. The median semimajor axis of the orbits of asteroids is 2.7 AU. Hence, the sidereal period of revolution of the median asteroid is about (a) 4.4; (b) 20; (c) 2.7; (d) 7.3 years.

736. If in fifteen weeks the line segment from the Sun to an asteroid sweeps out 0.1 of the total area enclosed by its orbit, about what is the period of revolution of the asteroid? (a) 29 years; (b) 2.9 years; (c) not calculable from the data given; (d) 4.3 years.

737. How can an asteroid be distinguished from a star with a simple optical telescope?

738. Which one of the following objects is an asteroid? (a) Europa; (b) Triton; (c) Juno; (d) Callisto.

739. Ceres is (a) an asteroid; (b) a satellite of Jupiter; (c) a satellite of Uranus; (d) a bright star in Cassiopeia.

740. Which one of the following is *not* an asteroid? (a) Hidalgo; (b) Europa; (c) Geographos; (d) Icarus.

741. Calculate the escape speed from the surface of a spherical asteroid whose mean density is 2.3 g cm^{-3} and whose radius is 5 km.

742. Would you expect an asteroid to have an atmosphere? Why or why not?

743. If you weigh 150 pounds on the Earth what would you weigh on the surface of an asteroid that has a mass $1/300,000^{th}$ that of the Earth and a radius $1/50^{th}$ that of the Earth?

744. The total energy emitted in the infrared by a particular asteroid is four times as great as the amount of sunlight it reflects. What is its optical albedo?

745. The Trojans are (a) satellites of Saturn; (b) a family of periodic comets; (c) a particular type of meteorite; (d) a special class of asteroids.

746. The Trojan asteroids (a) are in Keplerian orbits about Jupiter; (b) move in retrograde orbits between the orbits of Mars and Jupiter; (c) illustrate a special solution of the three-body gravitational problem; (d) have orbits that cross the orbit of the Earth.

747. The Trojan asteroids (a) are in Keplerian orbits about Jupiter; (b) are near the Lagrangian points L_4 and L_5 relative to the Sun and Jupiter; (c) are slowly escaping from the solar system; (d) have perihelion distances less than 1 AU, such that they cross the orbit of the Earth.

748. By what observational technique can the rotational period of an asteroid be determined?

749. The rotational periods of many asteroids have been determined by (a) observing cyclic departures from Keplerian orbits; (b) observing cyclic variations in their brightness; (c) supposing that they are tidally locked to the Sun as they revolve in their orbits; (d) observing the perturbations of their orbits by Jupiter's gravitational field.

750. About 65,000,000 years ago, dinosaurs and many other species of animal and plant life suddenly became extinct. Describe a current hypothesis for this effect.

751. What is the essential physical distinction between an asteroid and a comet?

752. According to Oort's theory, comets come from a cloud of icy bodies in Keplerian orbits about the Sun at a distance of the order of 50,000 AU. If such a body is perturbed from its orbit by a passing star so as to fall inwards toward the Sun, how long will it take to do so?

753. A comet that is perturbed from the Oort comet belt at 50,000 AU into an orbit with perihelion at 1 AU spends about (a) 11.2×10^6; (b) 2.0×10^6; (c) 5.6×10^6; (d) 4.0×10^6 years on its inbound flight to perihelion.

754. Concerning comets: (a) It is reasonably certain that most comets come from outside the solar system. (b) Kepler's laws are found to be useless in describing the motion of comets. (c) The orbit of a comet can be determined by observing its motion on the star field and using Kepler's laws. (d) The strength of the Sun's gravitational field is such that it is impossible for a comet to strike the Earth.

755. By photographing the head of a comet on the star field, its right ascension and declination can be found. If this is done at three successive times (e.g., at ten-day intervals) it is possible to determine its orbit relative to the Sun. What basic assumption is employed in order to make such a determination?

756. Most periodic comets (a) have retrograde orbits; (b) have aphelion distances greater than 3,000 AU; (c) have prograde orbits; (d) have passed through perihelion only once during recorded human history.

757. Concerning comets: (a) Kepler's laws show that it is impossible for a comet to strike the Earth. (b) Tables of the dates on which near-parabolic comets will appear are published in advance in *The Astronomical Almanac*. (c) It is reasonably certain that most comets come from outside the solar system. (d) The orbit of a comet can be determined by observing its motion on the star field.

758. The motion of periodic comets is dominated by the gravitational attraction of the Sun. Name another force that is often of significance.

759. Comets in near-parabolic orbits (a) have orbital periods of less than 40 years; (b) have usually passed close to the Earth as they approach the Sun; (c) have prograde orbits only slightly inclined to the ecliptic plane; (d) have orbits inclined at more-or-less random angles to the ecliptic plane.

760. A comet is observed to have a clearly hyperbolic orbit with respect to the Sun. Give two alternative hypotheses concerning the comet's previous orbital history.

761. Almost all of the known periodic comets (a) have aphelion distances greater than 100 AU; (b) have prograde orbits; (c) have retrograde orbits; (d) have randomly oriented orbital planes.

762. In order that a comet have an orbital period of less than 150 years, the semimajor axis of its orbit must be less than _____ AU.

763. By what natural process can the orbit of a comet be changed from a parabolic to an elliptical one or vice versa?

764. Why is the tail of a comet longer when near the Sun than when remote from it?

765. A comet usually develops a visible tail only within a radial distance of a few AU from the Sun. The principal cause of the tail is (a) solar heating of volatile constituents; (b) tidal forces by the Sun; (c) gravitational perturbations by Jupiter; (d) frictional heating by the interplanetary gas as the comet's speed increases near the Sun.

766. The brightness of a given comet head varies approximately as the inverse fourth power of its distance from the Sun (for constant distance from the comet to the Earth). Suggest an explanation for such a strong dependence.

767. The brightness B of an asteroid or a comet is given by the approximate formula

$$B \propto 1/R^n \Delta^2 \ ,$$

where R is the distance of the object from the Sun and Δ its distance from the Earth. For asteroids n is 2 whereas for comets n is typically 4. Explain this difference.

768. Suppose that the spherical nucleus of a comet 2 km in radius is completely disintegrated into spherical dust grains each 0.0001 cm in radius and that the resulting dust cloud is widely dispersed. By what factor will the amount of reflected sunlight have been increased?

769. The freezing or sublimation temperatures in degrees Kelvin of several common compounds are as follows: H_2O, 273; CH_4, 91; NH_3, 195; CO_2, 195. Given that the average radiative equilibrium temperature of the surface of the Earth is 270 degrees Kelvin and that this temperature of a body in the solar system varies as the inverse square root of the distance from the Sun, which of the above substances in the surface material of the nucleus of a comet would be in solid form at 4 AU?

770. Many comets have two different types of tails. What are the observational and physical distinctions between them?

771. Why does the gaseous tail of a comet extend radially outward from the Sun, irrespective of the direction of motion of the head of the comet?

772. The spectra of the dust tails of comets (a) are essentially identical to the spectrum of sunlight; (b) are dominated by the spectra of complex organic molecules; (c) reveal the presence of many radioactive elements; (d) often show the presence of sodium and iron.

773. The optical spectra of the coma of a comet (a) are identical to the spectrum of sunlight; (b) provide important information about the composition of cometary material; (c) reveal the presence of many radioactive elements; (d) are dominated by the emission spectra of complex organic molecules.

774. List four chemical elements whose presence in comets has been shown conclusively by spectroscopic data.

775. What is the evidence that comets contain ordinary water ice?

776. A comet at a distance of 1 AU from the Earth has a tail whose angular length is 10 degrees, as viewed at right angles to its length. Hence the length of the tail is about (a) 93,000,000; (b) 16,000,000; (c) 5,730,000; (d) 9,300,000 miles.

777. When near perihelion the tail of Comet Ikeya-Seki was observed in 1965 from 1 AU to have an angular length of about 20 degrees. What was its length in miles?

778. Which of the following names is identified with a particular comet? (a) Vesta; (b) Copernicus; (c) Ikeya-Seki; (d) Planck.

779. Comet Halley was predicted to pass through perihelion in early 1986. Name a conceivable reason, which could be checked, why this prediction might have failed markedly.

780. Comet Whipple has a sidereal period $P = 7.44$ years. Therefore, the semimajor axis of its elliptical orbit about the Sun (Kepler's third law) is (a) 20.29; (b) 55.34; (c) 1.94; (d) 3.81 AU.

781. The orbit of Comet Halley is an ellipse with one focus at the Sun, with an eccentricity of 0.97, and with a semimajor axis of 17.8 AU. Hence it can be expected to pass through perihelion at intervals of how many years? (a) 17.8; (b) 8.9; (c) 86.3; (d) 75.1.

782. Comet Halley has been observed since 239 B.C. to pass through the perihelion of its orbit at 76 year intervals. What is the semimajor axis of its orbit? (a) 8.9; (b) 18; (c) 36; (d) 663 AU.

783. Comet Halley has a period of 76 years and a perihelion distance of 0.6 AU. What is its approximate aphelion distance?

784. Comet Kohoutek was a much less spectacular object at perihelion passage than had been predicted soon after its discovery. The principal uncertainty in predicting the development of a comet is attributable to (a) inaccuracies in calculating its orbit; (b) unanticipated clouds of dust in interplanetary space; (c) lack of knowledge of its volatile constituents and physical structure; (d) sporadic variability in the intensity of sunlight.

785. Comet Kohoutek passed through the perihelion of its parabolic orbit on 28 December 1973 at a distance of 0.14 AU from the Sun. Using Kepler's laws, estimate its speed at that time. The orbital speed of the Earth is 30 km s^{-1}.

786. The common term "shooting star" refers to (a) a comet; (b) an asteroid; (c) a meteor; (d) a nova.

787. The best time of the night to observe bright meteors is during the early morning hours (about two hours before sunrise). The principal reason for this is that (a) there is less likelihood of moonlight in early morning; (b) the sky is usually clearer at these hours; (c) the Earth's orbital motion is such as to increase the relative speed of the Earth and a meteoroid; (d) the Earth's orbital motion is such as to decrease the relative speed of the Earth and a meteoroid.

788. Existing evidence supports the hypothesis that meteorites (a) are objects returning to the Earth after having been thrown off the Earth at an earlier time from erupting volcanoes; (b) come from Jupiter; (c) are small asteroids; (d) are the source of life on Earth.

789. Meteor showers (a) occur unpredictably; (b) are probably caused by cometary debris; (c) occur about once in five years on the average; (d) are caused by solar flares.

790. Show by a simple sketch why a particular meteor shower occurs on about the same date of each year.

791. The Perseids are (a) shower meteors; (b) a cluster of stars in the constellation Taurus; (c) a family of asteroids; (d) satellites of Neptune.

792. Meteors that appear in a shower each year in mid-August are called Perseids. What is the significance of this designation?

793. The meteors which occur in a shower each year around 14 December are called Geminids. What is the meaning of the term Geminids?

794. Describe an observational arrangement that will enable one to measure the velocity of a meteoroid in its passage through the atmosphere of the Earth.

795. Two cameras are located 50 km apart on an East-West line. A meteor trail is photographed by both cameras. The western camera shows that its trail began at the zenith while the eastern camera shows that its trail began at a point 55 degrees above the West point on the horizon. At what height above the Earth did the meteor trail begin?

796. Meteor trails at a height of 50 miles can be seen above the horizon of a sea level observer within a circle of about (a) 1,300; (b) 150; (c) 630; (d) 50 miles radius.

797. An observer reports seeing a meteor pass overhead at a height of about one mile. Criticize the credibility of this report.

798. In terms of their gross composition, meteorites are classified into two principal categories. What are these categories?

799. What are the essential steps in testing the claim that a particular object is a meteorite?

800. All *but one* of the following have been found in recovered meteorites: (a) crystallized minerals; (b) carbon compounds; (c) fossilized skeletons of primitive animals; (d) radioactive elements.

801. The most reliable determination of the age (lapse of time since solidification) of meteorites is yielded by radioactive dating. What is this age?

802. In a particular meteorite the abundance ratio $Pb^{206}/U^{238} = 0.41$. Assuming that there was no Pb^{206} in the meteorite at the time of its formation and noting that U^{238} decays to the stable nucleus Pb^{206} with a half-life of 4.5×10^9 years, find the age of the meteorite.

803. At the time of original solidification of a meteorite the atomic abundance ratios $Rb^{87}/Sr^{87}/Sr^{86}$ were 1.5/1.0/1.0. Rb^{87} decays with a half-life of 5×10^{10} years to produce Sr^{87}, whereas both Sr^{87} and Sr^{86} are stable nuclei and Sr^{86} is not the daughter of any radioactive nucleus in meteorites. Calculate the abundance ratios 5×10^9 years after solidification.

804. What speed with respect to the Earth may meteoroids be expected to have?

805. A 10-ton meteorite moving at typical speed has kinetic energy equal to the explosive energy of a 1-kiloton atomic bomb. Hence its impact with the Earth may be expected to (a) produce a large amount of radioactivity; (b) result in the vaporization of much of its mass; (c) cause a large increase (or decrease) in the rotational period of the Earth; (d) split the Earth into two or more pieces.

806. Name a prominent meteoric crater on the Earth.

807. There are far more known meteoric craters per unit area on the surface of the Moon, Mars, and Mercury than on the surface of the Earth. List plausible reasons for this great disparity.

808. About a thousand meteorites fall on the Earth each year. Assuming that their impacts are randomly distributed over the Earth's surface and that there are four billion persons on the Earth, estimate the probability that some person will be struck by a meteorite in one year.

CHAPTER 7

Radiations and Telescopes

809. Only one of the following is a form of electromagnetic radiation. Which one? (a) cosmic ray; (b) proton; (c) neutron; (d) gamma ray.

810. Which one of the following is *not* a type of electromagnetic radiation? (a) X ray; (b) ultraviolet; (c) cosmic ray; (d) gamma ray.

811. What are cosmic rays?

812. The first measurement of the speed of light was made by (a) Fizeau; (b) Roemer; (c) Michelson; (d) Galileo.

813. Describe the essential elements of the determination of the speed of light by observation of a satellite of Jupiter.

814. The apparent periods of revolution of the satellites of Jupiter vary cyclically during the course of a year. This effect is attributed to the (a) precession of their orbits; (b) tilt of the axis of the planet to the plane of its orbit; (c) finite speed of light; (d) mutual gravitational perturbations.

815. The true sidereal period of revolution of Io (one of the satellites of Jupiter) is 42^h 28^m. What is its apparent period to an observer on the Earth when the Earth is approaching Jupiter at a relative speed of 30 km s^{-1}?

816. What is the magnitude of the speed of light?

817. Describe an experimental method for measuring the speed of light.

818. What is the time of transit of a light pulse from the Sun to the Earth?

819. Noting that a light signal requires 8.32 minutes to travel from the Sun to the Earth, calculate the number of astronomical units in a light year.

820. From the fact that a light signal travels from the Sun to the Earth in about 8.3 minutes, find the ratio of the speed of light to the orbital speed of the Earth.

821. How does the intensity of light vary with distance from a point source?

822. The brightness of sunlight at the Earth is _____ times as great as at Pluto (40 AU from the Sun).

823. The intensity of sunlight at the Earth is _____ times as great as it is at Neptune (30 AU from the Sun).

824. What is the relative brightness of sunlight at the Earth and at Jupiter (5.2 AU from the Sun)?

825. A spacecraft at the orbit of Uranus (20 AU from the Sun) uses a parabolic mirror to collect sunlight for powering its electronic equipment. Assuming that 200 watts of electrical power are required and that the conversion of sunlight to electrical power is 10 percent efficient, find the necessary diameter of the mirror. At the Earth, 1,360 watts of sunlight strike one square meter.

826. A spacecraft at the orbit of Pluto (40 AU from the Sun) uses a parabolic mirror to collect sunlight for powering its electronic equipment. Assuming that 200 watts of electrical power are required and that the conversion of sunlight to electrical power is 10 percent efficient, find the necessary diameter of the mirror. At the Earth, 1,360 watts of sunlight strike one square meter.

827. Consider two identical stars A and B. A is four times as far away from the Earth as is B. What is the ratio of their apparent brightnesses?

828. At perihelion the planet Mercury is 0.31 AU from the Sun. How much brighter is sunlight there than at the Earth?

829. When the spacecraft Pioneer 10 was 52 AU from the Sun, its instruments were at a temperature of 260 degrees Kelvin; but Saturn at 10 AU has a surface temperature of only 75 degrees Kelvin. How could the Pioneer 10 instruments be so warm?

830. The temperature of an inert object in interplanetary space is dependent primarily on (a) the amount of sunlight that it absorbs; (b) its speed through the interplanetary medium; (c) the temperature of the gas in the solar wind; (d) the amount of dust with which it collides.

831. The radiative equilibrium temperature of an inert body in space varies directly as the fourth root of the amount of radiation that it absorbs from external sources. If the intensity of starlight is 10^{-8} of that of sunlight at 1 AU, what would be the temperature of a body in "deep space" (very remote from the Sun)?

832. Barnard's star has a proper motion of 10 seconds of arc per year and is at a distance of 1.8 parsecs (pc). What is the component of its space velocity perpendicular to the line of sight, in km s^{-1}? Use the approximations 1 pc $= 3 \times 10^{13}$ km and 1 year $= 3 \times 10^7$ s.

833. Einstein's special theory of relativity (a) is an interesting hypothesis but has no relationship to known phenomena; (b) predicts that the universe will expand for a few billion years and will then start contracting; (c) assumes that the speed of light is independent of the relative motion of the source and the observer; (d) gives a basis for the concept of absolute motion.

834. *All but one* of the following conceivably observable effects are in accordance with Einstein's special theory of relativity. (a) The mass of an electron increases as its speed increases. (b) The decay lifetime of a radioactive particle increases as its speed increases. (c) The speed of light is the same regardless of whether an observer is moving toward or away from the source of light. (d) The laws of physics are quite different in any coordinate system that is in motion with respect to the Earth.

835. If an object moves with a speed v′ in a system which itself is moving with a speed V with respect to observer O, and if the direction of v′ is parallel to that of V and in the same sense, the object's speed with respect to O is given by the relativistic formula

$$v = \frac{v' + V}{1 + \dfrac{v'V}{c^2}}$$

where c is the speed of light. If v′ = V = 0.8 c, what is v?

836. Two objects are moving along a line toward each other, each with speed 0.6 c in a particular observer's coordinate system. According to Galileo their relative speed is 1.2 c. According to Einstein it is (a) 1.0 c; (b) 0.6 c; (c) 1.2 c; (d) 0.88 c.

837. The wavelength of the electromagnetic waves from a television station operating at a frequency of 75 megahertz is (a) 4 meters; (b) 25 cm; (c) 2.5×10^{-3} cm; (d) 400 meters.

838. A weather radio station operates on a frequency of 163 megahertz. What is the wavelength of the electromagnetic waves that it broadcasts?

839. A radio station broadcasts on a frequency of 910 kilohertz (i.e., 9.10×10^5 cycles per second). What is the wavelength of its radiated signals?

840. A typical television station transmits at a frequency of 100 megahertz $(= 1 \times 10^8$ cycles per second). What is the wavelength of its electromagnetic radiation?

841. Describe a spectroscopic method for finding the radial speed of a distant star relative to the solar system.

842. The Doppler effect has been used for determination of the (a) surface composition of planets; (b) temperature of the Sun; (c) diameter of Pluto; (d) rotational period of Mercury.

843. An observer, equipped with a mercury arc, photographs the spectrum of the arc after reflection of the light by a mirror that is approaching at a speed of 30 km s^{-1}. The Doppler shift of the 5461 angstrom line of the mercury spectrum is (a) 0.55 angstrom toward the red; (b) 0.55 angstrom toward the violet; (c) 5.46 angstroms toward the red; (d) 1.09 angstroms toward the violet.

844. A stellar spectral line whose laboratory wavelength is 5,800 angstroms is shifted by the Doppler effect toward the red by 0.29 angstrom. Hence, the radial speed of the star relative to the observer is _____ km s^{-1}
(a) toward the observer.
} Select one.
(b) away from the observer.

845. The Doppler effect is useful in astronomy for measuring (a) the angular separation of two planets; (b) the temperature of a star; (c) the radial speed of a star; (d) the diameter of a star.

846. The Bohr theory of the hydrogen atom envisions one electron in a Keplerian orbit about the proton (nucleus of the atom). Apart from differences in the scale of distances and the nature of the forces involved, what is the basic difference between this model and the case of a planet orbiting the Sun?

847. An atom can be ionized by *all but one* of the following:
(a) a strong electrical field; (b) collision with another atom;
(c) X rays; (d) a uniform gravitational field.

848. What physically measurable property of a beam of monochromatic light identifies its color? (a) intensity; (b) wavelength;
(c) speed of propagation in vacuo; (d) line breadth.

849. The dispersion of light into its constituent colors (a spectrum)
by a ruled grating depends upon the physical principle called
(a) refraction; (b) diffraction; (c) reflection; (d) absorption.

850. The bending of a beam of light at an interface between glass
and air is called (a) aberration; (b) parallax; (c) refraction;
(d) diffraction.

851. Upon what optical principle does image formation by a lens
depend?

852. Explain how a glass prism can be used to examine the spectrum of light from a star.

853. The basic optical properties of a simple lens depend upon the
phenomenon of (a) reflection; (b) diffraction; (c) refraction;
(d) dispersion.

854. After passage of a beam of white light through a glass prism the
order in which the colors appear in the resulting spectrum—
from least deviated to most deviated (G = green, V = violet,
R = red, Y = yellow)—is (a) GVRY; (b) VYGR; (c) RGYV;
(d) RYGV.

855. The dispersion of light into its constituent colors (a spectrum)
by a glass prism depends upon the physical principle called
(a) refraction; (b) absorption; (c) diffraction; (d) reflection.

856. A diffraction grating is useful for (a) correcting spherical aberration of a mirror; (b) correcting chromatic aberration of a lens; (c) examining the spectrum of starlight; (d) magnifying an image.

857. A spectrograph is used in astronomy for (a) measuring the Doppler effect; (b) finding the distance to a planet; (c) finding the size of lunar craters; (d) determining the equation of time.

858. A device called a (a) spectroheliograph; (b) micro-densitometer; (c) polarimeter; (d) coronagraph is used for producing artificial eclipses of the Sun.

859. A waterproof camera placed under water and pointed vertically upward has a field of view that includes the entire hemisphere above the surface of the water (a "fish-eye" camera). Show by a ray-tracing diagram how this can be true.

860. An illuminated object is placed two meters away from a spherical concave mirror of 0.5 meter focal length. Its image is (a) virtual and erect; (b) real and inverted; (c) real and erect; (d) virtual and inverted.

861. In the optical system shown below, F is the focal point of the concave parabolic mirror MM′, and AB is an illuminated object. Find the orientation, size, and approximate position of the image A′B′ by ray tracing.

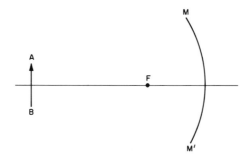

862. MM′ represents the surface of a convex spherical mirror whose center of curvature is at O. A light ray strikes the mirror at P. Draw the reflected ray and explain how you determined its direction.

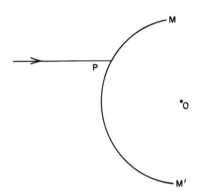

863. An optical system consists of two converging lenses and a plane mirror as shown. An illuminated arrow AB is placed in the focal plane of the first lens. Find the position and orientation of its image.

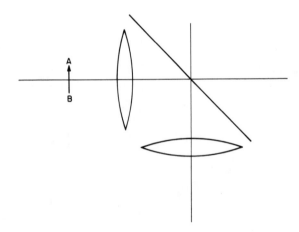

864. The images formed by a convex spherical mirror are always (a) inverted; (b) virtual; (c) real; (d) enlarged.

865. An illuminated arrow AB (the object) is placed in front of a concave spherical mirror with center of curvature at O at a distance from the mirror of twice its radius of curvature. Construct the image by ray tracing and then make the proper choice between the two possibilities in each of the following cases. The image is
(a) erect or inverted,
(b) larger or smaller than the object,
(c) real or virtual.

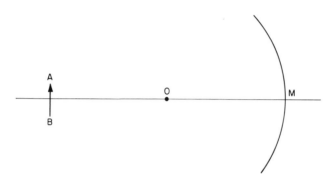

866. In the optical system shown below, F, F' are the focal points of the converging lens and AB is an illuminated object. Find the approximate position and orientation of the image A'B' by ray tracing.

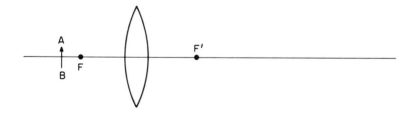

867. A particular camera has a 105 mm focal length lens. At what distance away must a two-meter tall person stand in order that the image will just span a "35 mm" film (assume 25 mm available)? (a) 16.8; (b) 10.5; (c) 8.4; (d) 4.2 meters.

868. Suppose that your eyes are 3 inches apart and that you sight over your forefinger which is held at a distance of 24 inches from your eyes. In viewing the star field, you see star A aligned with your finger when using your left eye, and star B when using your right eye. The angular separation of A and B is (a) $1/8^{th}$; (b) 12; (c) 7; (d) 4 degrees.

869. A person with good eyesight can resolve (i.e., discern as separate) two small lights one inch apart at a distance of 300 feet. What is the angle subtended by these two objects at the person's eye?

870. The resolving power of the human eye is about one minute of arc. Mention some common object that would subtend about this angle at a distance of 100 meters.

871. The resolving power of a telescope is increased by (a) increasing its focal length; (b) increasing the diameter of its objective; (c) working with red light only; (d) decreasing the diameter of its objective.

872. The maximum aperture of the human eye has a diameter of 1/4 inch and that of a particular telescope is 84 inches. Compare the light-gathering powers of the two instruments.

873. The narrow angle telescope on Voyager 1 has a focal length of 1,500 mm. At what distance from Saturn (diameter 120,000 km) would the image of the planet have a diameter of 10 mm?

874. The gaseous density and therefore the index of refraction of the atmosphere of a planet diminishes with increasing height. Hence, it is conceivable that a thin beam of light might be bent in such a way that it has a circular path around the planet. Find the necessary conditions.

875. The amount of starlight collected by an astronomical telescope is proportional to (a) the square of the diameter of its objective; (b) the f number of its objective; (c) the square of the focal length of its objective; (d) the ratio of the diameter to the focal length of its objective.

876. A particular telescope is said to be a 24-inch telescope. What dimension is it that is 24 inches?

877. What is the diameter (in inches) of the image of the Moon as photographed with a 24-inch telescope having a focal ratio of f/8?

878. The light-gathering power of a 25-inch telescope is about (a) 1,000; (b) 100; (c) 100,000; (d) 10,000 times greater than that of the human eye. Assume that the diameter of the entrance aperture of the dark adapted eye is 1/4 inch.

879. The University of Iowa's Cassegrain telescope near Hills has an effective focal length of 490 cm. What is the diameter of the image of Venus when its distance is such that its angular diameter is 30 seconds of arc? (a) 0.07; (b) 7; (c) 4.2; (d) 0.7 mm.

880. Suppose that one wishes the Moon to appear 25 degrees across when viewed through a telescope of one meter focal length. What focal length eyepiece must be used? (a) 10; (b) 2; (c) 1; (d) 5 cm.

881. Compare the properties of two telescopes
    X (F = 32 ft, d = 24 in) and
    Y (F = 6 ft, d = 6 in)
for photographing a particular portion of the Moon (F = focal length, d = diameter of objective). (a) X is faster than Y; (b) Y is faster than X; (c) Y produces a larger image than X; (d) Y has better resolution than X.

882. A parallel beam of light in air strikes a solid uniform sphere of transparent refractive material having index of refraction n and radius a. It is readily shown that the focal point of rays having different angles of incidence i to the normal to the spherical surface varies with i for a given n. Find the value of n such that a circular ring of parallel rays all having a given i is focussed at a point on the surface of the sphere.

883. A beam of light in glass (index of refraction 1.5) strikes the plane glass-air interface at an angle of incidence i. Find the range of the values of i for which the beam is totally reflected and does not penetrate the interface.

884. The McMath solar telescope at the Kitt Peak National Observatory has a focal length of 300 feet. The diameter of the image of the Sun at the prime focus is (a) 6.48 inches; (b) 33 inches; (c) 5.0 feet; (d) 44.8 cm.

885. The angular diameter of the Moon is 30 minutes of arc. What is the diameter (in inches) of the image of the Moon formed by a telescope having a focal length of six feet?

886. The diameter of the image of the Moon in the focal plane of a particular telescope is 1 cm. What is the focal length of the telescope?

887. Compare the brightnesses of images of the Moon formed at the prime focus of telescope A whose objective mirror has a diameter of 50 inches and a focal length of 400 inches and telescope B whose objective mirror has a diameter of 10 inches and a focal length of 40 inches.

888. Using a telescope on the Earth it is seldom possible to distinguish two objects on the Moon that are closer together than about (a) 100 feet; (b) 5 miles; (c) 20 miles; (d) 10 miles; (e) 1 mile.

889. A 24-inch Cassegrain telescope has an effective focal length of 384 inches. A properly exposed photograph of the Moon's surface can be obtained on a particular type of film in one second. Hence, if one uses the same type of film in an ordinary personal camera set at f/1.6, a properly exposed photograph of the Moon can be obtained in (a) 1/100th; (b) 1/10th; (c) 100; (d) 10 s.

890. Compare the observational characteristics of two telescopes X and Y, both of which have "perfect" optical properties.
X: 24-inch diameter objective, f/8.
Y: 96-inch diameter objective, f/16.

891. The objective lens of a five-inch refractor has a focal length of 6 feet. The magnification of this telescope with an eyepiece of 1/2 inch focal length is (a) 144; (b) 3; (c) 12; (d) 72.

892. If the Moon can be photographed in one-hundreth of a second with a telescope of 10-inch aperture and 40-inch focal length, how long a time would be required to photograph the Moon with a telescope of 20-inch aperture and 200-inch focal length?

893. A two-inch diameter hole is bored through the center of the eight-inch diameter objective mirror of a telescope. As a result (a) the resolving power of the telescope is reduced by about 25 percent; (b) the telescope is useless for viewing stars lying along its axis; (c) the light-gathering power of the telescope is reduced by about 6 percent; (d) an image of the Moon will have a dark spot in its center.

894. A "corner reflector" consists of three mutually orthogonal mirrors. Prove that a ray of light that enters such a device will, after a total of one, two, or three reflections, be reversed in direction by exactly 180 degrees.

895. The light-gathering power of a 200-inch telescope is (a) 64; (b) 16; (c) 8; (d) 32 times as great as that of a 25-inch telescope.

896. The Moon is photographed with telescope A whose objective has a diameter of 10 inches and a focal length of 5 feet, and then with telescope B whose objective has a diameter of 200 inches and a focal length of 40 feet. Find (a) the ratio of the appropriate exposure times and (b) the diameters of the images in inches on the photographic plate in the two cases.

897. In using eyepiece projection of the image of the Sun with a five-inch refractor the image of the Sun on the screen is (a) erect and larger; (b) inverted and larger; (c) erect and smaller; (d) inverted and smaller than the image formed by the objective.

898. In order to photograph faint stars in the presence of a general luminous glow in the sky, it is advantageous to select a telescope of the (a) largest available f number; (b) smallest available f number; (c) largest available aperture; (d) largest available focal length.

899. Which one of the following statements is *false*? (a) The prime focus of a telescope is never accessible to observing equipment. (b) The Coude optical arrangement is advantageous when using a large spectrograph. (c) It is possible to view all objects above the horizon with a fixed vertical telescope above which is a swiveled plane mirror. (d) Spherical aberration in a reflecting telescope is sometimes reduced by a glass-correcting plate.

900. The principal advantage of an equatorial-polar mount over an altitude-azimuth mount for an astronomical telescope is that (a) an astronomical object can be tracked by rotating the mount about only one axis; (b) it is less expensive to build; (c) circumpolar stars cannot be observed with the latter; (d) it is easier to align accurately.

901. An astronomical telescope is mounted on a polar-equatorial mount and equipped with a motor drive such that the axis of the telescope is kept pointed continuously at the Moon. In order to do this, the rotational rate in hour angle must be 360 degrees in about (a) $23^h 56^m$; (b) $23^h 10^m$; (c) $24^h 04^m$; (d) $24^h 50^m$.

902. The 200-inch Hale Telescope at Mount Palomar (A) has a focal length of 660 inches (prime focus). A particular camera (B) has a lens of 1-inch diameter and 1.4-inch focal length. In photographing Jupiter on the same type film, the proper exposure time with A is _____ times that with B.

903. The best astronomical seeing through the atmosphere of the Earth permits a resolution of one second of arc. What is the minimum separation of two features on the surface of the Moon (in kilometers) such that they can be just distinguished from each other by a ground-based telescope? The distance to the Moon = 384,400 km.

904. An earthward-pointed camera having a high quality six-inch diameter objective is carried by a reconnaissance satellite at a height of 100 miles. Ideally two point sources of light at the following separation can be distinguished (i.e., resolved): (a) 100; (b) 20; (c) 50; (d) 2 ft.

905. Sketch the optical system of a Cassegrain reflecting telescope showing positions, forms (plane, spherical, hyperbolic, parabolic, etc.) and types (lens, mirrors, prisms, etc.) of all essential optical elements. Where is the image?

906. The practical limit on the resolving power of conventional ground-based optical telescopes is attributable to (a) diffraction; (b) atmospheric shimmer; (c) seismic vibration of the mount; (d) difficulty of making accurate optical surfaces of large dimensions.

907. A reflector having an aperture of five feet has a one-foot diameter hole in the center of the objective mirror. The principal effect of this hole is to (a) increase spherical aberration slightly; (b) make it impossible to image objects that are near the optical axis; (c) reduce the brightness of stellar images by four percent.

908. Encircle *one or more* of the following claims as *true* for an optical telescope: (a) The larger the objective, the greater is the resolving power. (b) The greater the focal length, the larger is the primary image. (c) The larger the objective, the larger is the primary image. (d) The greater the focal length, the greater is the light-gathering power. (e) The larger the objective, the brighter is the primary image.

909. What advantages are there in placing an astronomical optical telescope above the atmosphere of the Earth?

910. The most fundamental advantage of a stellar telescope in a satellite over one on the ground is (a) the freedom from vibration; (b) the absence of clouds; (c) the accessibility of the ultraviolet portion of the spectrum; (d) the elimination of atmospheric shimmer.

911. The spectrum of a particular light beam looks like this sketch. The source is: (a) a neon sign; (b) a hot solid (or hot, dense gas); (c) a sodium flame; (d) a hot, tenuous gas.

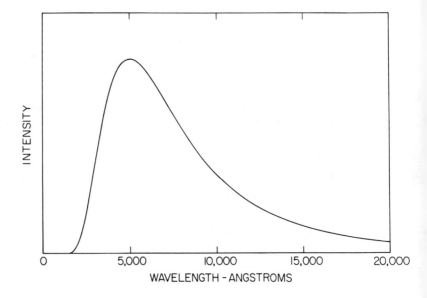

912. The optical spectrum of a given source looks something like the following sketch. The source is (a) a hot, tenuous gas; (b) a hot solid (or hot, dense gas); (c) a cool solid; (d) a black body.

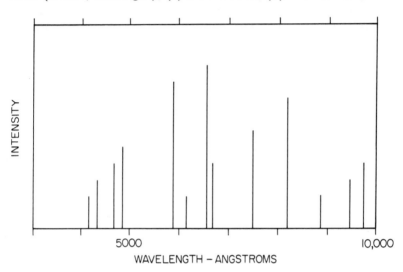

INTENSITY

5000

10,000

WAVELENGTH – ANGSTROMS

913. The spectral intensity of light from a particular star has its greatest value at a wavelength of 4,500 angstroms (in the blue). On the basis of this evidence alone, it can be concluded that the star is (a) cooler than the Sun; (b) larger than the Sun; (c) at the same temperature as the Sun; (d) hotter than the Sun.

914. The spectrum of sunlight reflected from a planet provides information on the gaseous composition of the atmosphere of the planet. Give a brief explanation.

915. Chemical elements in the Sun's atmosphere can be identified by (a) the Doppler effect; (b) measuring the temperature of the Sun's disc; (c) their characteristic absorption lines in the solar spectrum; (d) observing the color of the Sun through the Earth's atmosphere near sunset.

916. The maximum spectral intensity of a star occurs at a wavelength of 8,000 angstroms. Therefore, the effective temperature of the star is _____.

917. Several months after the appearance of Nova Cygni 1975 its spectrum was dominated by emission lines. Hence (a) it was receding from the Earth at high speed; (b) its atmosphere consisted principally of hot, tenuous gas; (c) its atmosphere consisted principally of hot, very dense gas; (d) its atmosphere consisted principally of cool gas surrounding a hot solid core.

918. What is a radio telescope?

919. Radio telescopes commonly use paraboloidal-shaped mirrors ("dishes"). Explain why such a shape is superior to a spherical one.

920. One radio telescope is 60 feet in diameter and the other, 28 feet in diameter. The larger collects about (a) 2.1; (b) 10; (c) 4.6; (d) 3.7 times as much radio power as does the smaller.

921. The resolving power of a 20-meter diameter radio telescope at 150 megahertz is about (a) six minutes of arc; (b) two degrees of arc; (c) twelve degrees of arc; (d) six degrees of arc.

922. What is the angular resolution of a 70-meter diameter radio telescope operating at 2,000 megahertz?

923. A radio interferometer is physically analogous to (a) a refracting telescope; (b) a prism; (c) an optical spectrometer; (d) an optical diffraction grating.

924. Give a brief list of the contributions of radio and radar techniques to lunar and planetary astronomy.

☆ ☆ ☆

ANSWERS TO

CHAPTER 1

The Sun and the Nature
of the Solar System

1. Consider: The orbital planes and senses of revolution of the planets and asteroids; the sense of rotation of the Sun; the sense of revolution of planetary satellites and ring material; the composition and density of planets as a function of heliocentric distance; the large number of asteroids and their orbits; the cratering of the solid surfaces of the Moon, Mercury, Mars, Venus, asteroids, and planetary satellites; and the radiogenic ages of the Earth, the Moon, and meteorites.

2. (d) Kuiper.

3. (a) Were developed by accretion from a primordial cloud of gas and dust.

4. (b) The Sun will have about its present size and brightness for at least another billion years.

5. (b) Elements such as iron, nickel, chromium, and molybdenum are concentrated near the surface. (False)

6. (a) Is the largest object in the solar system.

7. (c) That there probably are planetary systems associated with many millions of other stars.

8. (d) Nuclear fusion of hydrogen to form helium.

9. $(4\pi r^2)(1.4) = 4 \times 10^{23}$ kilowatts.

10. Mean density $= 1.41$ g cm^{-3} plus equation of state of material under self-gravitational pressure.

11. (c) The photosphere.

12. (d) 5,800 degrees Kelvin.

13. (c) In the blue-green portion of the spectrum.

14. (d) Absorption lines of many different elements in the spectrum of the whole disc of the Sun.

15. Identification of Fraunhofer lines with laboratory spectra.

16. (a) Absorption lines in the spectrum of photospheric light.

17. Density $\rho = \rho_0 (1 - r)$ for $0 \le r \le 1.0$. Integrate dm $= 4\pi r^2 \rho$ dr from r $= 0$ and r $= 1$ to find the total mass m $= \pi \rho_0 /3$, then integrate dm from r $= 0$ to such an upper limit as to give $\pi \rho_0 /6$. Result r $= 0.614$.

18. 270,200 AU (1.31 parsec).

19. About one-half of the distance to the nearest star, Proxima Centauri, that is, 135,100 AU.

20. 215.

21. (b) 330,000 times as great as that of the Earth.

22. (b) A spherical shell of gas about 3,000 km thick.

23. 6.7% (assuming 5×10^9 as the age of the solar system).

24. (a) About 1 million degrees Kelvin.

25. (a) Magnetized dust. (False)

26. By observing the day-to-day east to west movement of durable sunspots across the visible disc of the Sun.

27. Use eyepiece projection of the image of the Sun onto a white screen and observe the day-to-day motion of durable sunspots across the visible disc.

28. (a) 27 days.

29. (a) 11 years.

30. (b) Changes in the orbital motion of the Earth. (False)

31. (d) Factors of ten or more.

32. (b) Consists of hot ionized gas, mostly hydrogen, moving radially outwards from the Sun's corona.

Motion of the Planets

33. Because of their apparent motion relative to the star field.

34. Because their positions in the star field change continuously with time.

35. Any three of the following twelve Zodiacal constellations: Aries, Taurus, Gemini, Cancer, Leo, Virgo, Libra, Scorpius, Sagittarius, Capricornus, Aquarius, and Pisces.

36. Note observed date, time, and position on the star field and brightness.

37. (a) Copernicus.

38. Both concepts are purely geometric in nature. The Ptolemaic hypothesis (geocentric) in its simplest form envisions the uniform motion of a planet or the Sun along a circular orbit called an epicycle whose center in turn moves uniformly along a larger circle called the deferent centered on the Earth. The combination of epicycle and deferent is different for each body.

The diagram gives a schematic view of this concept. Note that the centers of the epicycles of Mercury and Venus are attached to a radial line from the Earth to the Sun and that no epicycle is shown for the Sun.

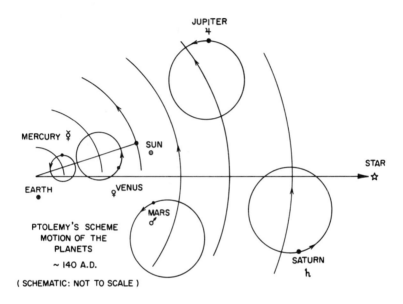

The Copernican hypothesis (heliocentric) in its simplest form envisions the uniform motion of the Earth and all of the other planets along circles of various radii centered on the Sun. Both schemes (more elaborate in detail) undertake to describe the observed motion of the planets and the Sun on the star field.

39. It is totally incompatible with Newton's law of gravitation and his three laws of motion.

40. The motion of planets and the Sun along the epicycles and deferents of Ptolemy's scheme has no relationship whatever to the masses, forces, and accelerations that are the heart of physical reality as recognized by Galileo and Newton.

41. The heliocentric hypothesis has at least a general resemblance to the physical reality of forces, masses, and accelerations whereas the geocentric hypothesis has no such resemblance.

42.

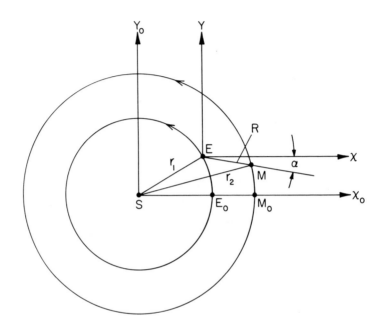

■ The first diagram depicts the Copernican scheme. Suppose that the initial configuration is an opposition with the Sun (S), the Earth (E_0) and Mars (M_0) aligned as shown along the X_0-axis. After a lapse of time t, the Earth is at E and Mars at M. A translated coordinate system XY is centered at E with the X-axis parallel to the X_0-axis.

By a sequence of carefully scaled drawings, the Earth-Mars distance R and the geocentric longitude of Mars α can be tabulated as a function of t and then assembled as a polar plot in the XY coordinate system.

The analytical solution is as follows with t in years:

$r_1 = SE = 1.0$
Angle E_0 SE $= 360\ t$
$r_2 = SM = 1.59$
Angle M_0 SM $= 180\ t$

Coordinates of M:

$$x = 1.59 \cos 180\, t - 1.0 \cos 360\, t$$
$$y = 1.59 \sin 180\, t - 1.0 \sin 360\, t$$
$$R = \sqrt{x^2 + y^2}$$
$$\sin \alpha = y/R$$
$$\cos \alpha = x/R$$

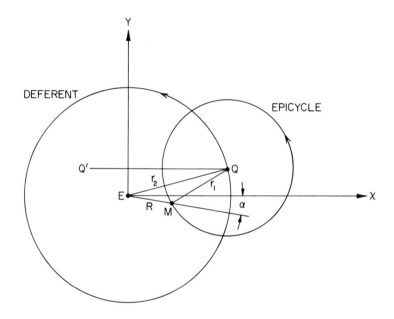

■ The second diagram depicts the Ptolemaic scheme. Q is the center of the epicycle, a circle of radius r_1. Q moves along the deferent, a circle of radius r_2, such that the angle QEX = 180 t. Mars is located on the epicycle and moves along its circumference so that angle Q'QM = 360 t. The X-Y coordinate system is centered on the Earth and the X-axis points at a distant star. QQ' is parallel to the X-axis. The initial configuration at t = 0 is with Mars and the center of the epicycle on the +X-axis. The diagram shows the configuration after a lapse of time t.

Again it is evident that a sequence of carefully scaled drawings will yield a tabulation of R and α as a function of time t.

The analytical solution is as follows:

Comparing the first and second diagrams, one notes (1) that angle ESM in the first diagram is equal to angle EQM in the second; (2) that SE in the first equals QM in the second, each being r_1; and (3) that SM in the first equals EQ in the second, each being r_2.

Hence, triangle ESM in the first diagram is identical to triangle MQE in the second diagram. It follows that M has the same x,y coordinates in the second diagram as it had in the first and hence the values of R and α as a function of time are identical for both the Copernican and Ptolemaic schemes in this illustrative example.   Q.E.D.

■ The third diagram shows the geocentric orbit of Mars for either case.

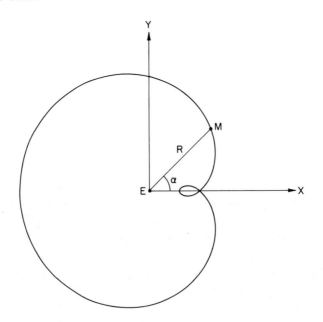

43. (b) Support the Copernican hypothesis.

44. Geometric visualization of a heliocentric scheme for motion of the planets, which presaged development of the physical theory of such motion.

The Ptolemaic scheme of planetary motion corresponded to the perception of ordinary experience that the Earth is stationary and that all celestial objects move with respect to it. It also incorporated the philosophical/religious (homocentric) concept that the Earth as the habitat of the human race had a unique status as the center of the universe.

Copernicus' advocacy of the heliocentric scheme of planetary motion relegated the Earth to minor astronomical status and, in the broad context of human thought, was both revolutionary and heretical.

45. (c) His observations of the apparent motion of the planets on the star field.

46. See answers to 40, 41, 42, 43, and 44.

47. (d) Kepler.

48. (c) From study of Tycho Brahe's observational data.

49. (b) Tycho Brahe's observations of the apparent motion of planets on the star field.

50. First Law: The orbit of each planet lies in a plane passing through the Sun and is an ellipse with one focus at the Sun.

    Second Law: The radial line segment from the Sun to a planet sweeps out equal areas per unit time as the planet moves along its elliptical orbit.

    Third Law: The square of the orbital period of revolution of a planet is proportional to the cube of the semimajor axis of its elliptical orbit.

51. Kepler identified the Sun as the central body in the motion of all planets, in general accord with the Copernican model but totally contrary to the Ptolemaic model.

52. P = 64.

53. Establish the positions of two points F_1 and F_2. Trace the locus of a point P such that $F_1P + F_2P = $ a constant (greater than F_1F_2). Note that $F_1P + F_2P = 2a$, where a is the semimajor axis of the ellipse.

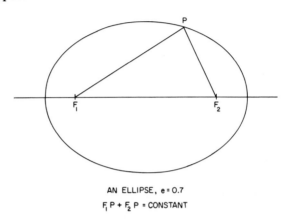

AN ELLIPSE, e = 0.7

$F_1 P + F_2 P = $ CONSTANT

54.

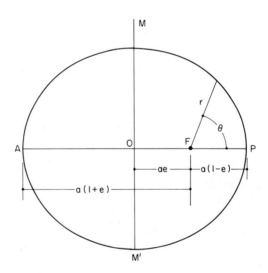

The diagram shows an example of the geometry of an ellipse having eccentricity e = 0.5. O is the center of the ellipse; the semimajor axis a = OA = OP; and the semiminor axis b = OM = OM′. AP is the line of apsides, P being the periapsis and A, the apoapsis. F is one of the two symmetrically located foci along the major axis. In the context of planetary motion F is the occupied focus (the Sun). The polar equation of the ellipse is

$$r = \frac{a(1 - e^2)}{1 + e \cos \theta}.$$

A full description of an elliptical planetary orbit in space also requires specification of the orientation of the plane of the orbit (two parameters) and the orientation of its line of apsides (one additional parameter) as well as specification of the epoch at which the planet passes through periapsis. Thus a total of six independent parameters are required. The rate of motion of the planet along its orbit is governed by conservation of angular momentum about F (Kepler's second law) but an explicit formula for the rate of motion is complex.

All six orbital parameters are uniquely derivable for injection of a planet into the Sun's gravitational field with a specified velocity vector (three components) at a specified position (three coordinates).

55. Nothing.

56. (a) Mass of an object is the quantity of matter therein, a scalar having magnitude only.
(b) Weight is the vector force on an object within the gravitational field of other objects. Its magnitude is proportional to the mass of the object.
(c) Velocity of an object is the instantaneous rate of change of its position. It is a vector quantity, having magnitude, direction, and sense.
(d) Speed, a scalar, is the magnitude of a velocity vector.
(e) Acceleration is the instantaneous rate of change of the velocity vector of an object. It is also a vector quantity, having magnitude, direction, and sense.

57. (a) 30 miles per hour.
(b) 21.21 miles per hour in the northwesterly direction.
(c) 10.61 miles per hour^2 in the northeasterly direction.

58. ∎ Accelerator pedal, used to increase speed.
    ∎ Brake pedal, used to reduce speed.
    ∎ Steering wheel, used to change direction of motion.

59. (c) Newton's first law of motion is valid.

60. (a) Newton's first law of motion is valid.

61.

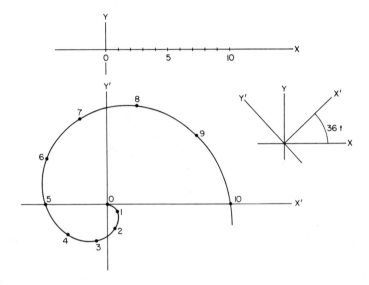

The ticks along the X-axis in the upper panel show successive positions of the object in the XY inertial coordinate system. The dots along the spiral in the lower panel show the corresponding positions in the rotating X'Y' system.

62. (b) Becomes greater as they are brought closer to each other.

63. (d) $F = G \dfrac{m_1 m_2}{r^2}$

64. $F = G \dfrac{m_1 m_2}{r^2}$

where F is the magnitude of the attractive force between two point masses m_1 and m_2 separated by a distance r. G is Newton's universal gravitational constant, the same for any material. The vector force on m_1 due to m_2 is directed toward m_2 and lies along the line joining m_1 and m_2. The vector force on m_2 due to m_1 is of equal magnitude, is directed toward m_1 and lies along the line joining m_2 and m_1.

65. These important theorems can be proved by direct integration of Newton's law of gravitation between two masses of infinitesimal dimensions. Such a procedure involves the integration of complicated transcendental functions. By way of contrast, Newton's original proofs, as given in his *Principia*, employ only elementary geometry and also give superior insight into the physics of the problems. The following proofs are adaptations of Newton's:

(a) Consider the force on a unit mass placed at an arbitrary point P *interior* to a thin spherical shell with center at O, radius a, and uniform mass per unit area m; and take G = 1. Draw an arbitrary chord through P, intersecting the shell at Q and Q' as shown in the first diagram

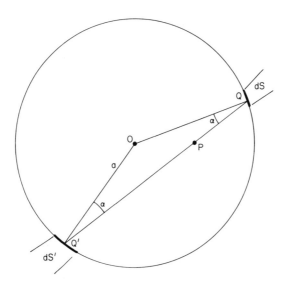

Then imagine equal cones of infinitesimal solid angle $d\Omega$ from P to Q with vertex at P and from P to Q' with vertex also at P, cutting out areas dS at Q and dS' at Q'. Angles OQP and OQ'P are equal and are denoted by α. The gravitational force on a unit mass at P by mdS is $mdS/(PQ)^2$ and by mdS' is $mdS'/(PQ')^2$. The two forces are oppositely directed. Further,

$$dS \cos\alpha/(PQ)^2 = d\Omega \quad \text{and} \quad dS' \cos\alpha/(PQ')^2 = d\Omega \ .$$

Hence the two forces are exactly cancelatory, being equal in magnitude and opposite in direction.

The entire shell can be so subdivided and because every pair of elements yields a zero resultant at the arbitrary point P, the entire shell exerts no force on a unit mass at any interior point P. This result is itself an important theorem and is obviously applicable to a thick spherical shell surrounding P, provided that the mass density of each concentric subshell is a function only of radius.   Q.E.D.

(b) Next consider the gravitational force of the previously specified thin spherical shell on a unit mass at an arbitrary *external* point P' as shown in the second diagram.

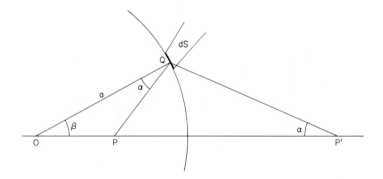

For any external point P′ there is an interior point P on the line OP′ such that $(OP)(OP') = a^2$. Q is an arbitrary point on the shell. The relationship between triangles OQP′ and OPQ is shown in the third diagram.

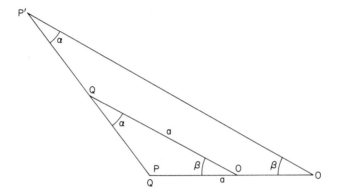

The external apex labels are for OQP′, and the interior apex labels are for OPQ. As shown in the second diagram the two triangles have a common angle β. Also, the special selection of point P assures that

$$\frac{OP}{OQ} = \frac{OQ}{OP'}$$

because $OQ = a$. Hence the two triangles are similar and the angles PQO and QP′O are equal to each other and each is equal to α, as shown.

Next imagine a cone of infinitesimal solid angle $d\Omega$ from P to Q with vertex at P. It cuts out an area dS of the shell at Q. The gravitational force exerted on a unit mass at the external point P′ is

$$\frac{m\,dS}{(QP')^2} .$$

By symmetry around the line OPP′, the resultant force exerted by the entire shell on a unit mass at P′ lies along this line. The contribution to this resultant force by dS is

$$m\cos\alpha\,\frac{dS}{(QP')^2} .$$

Further, by virtue of the similarity of the two triangles exhibited in the third diagram

$$\frac{QP'}{OP'} = \frac{PQ}{OQ}$$

or

$$QP' = \frac{(OP')(PQ)}{a} .$$

Hence the contribution by dS to the resultant force at P' becomes

$$m\,a^2 \cos\alpha \, \frac{dS}{(OP')^2(PQ)^2}$$

and because

$$\cos\alpha \, \frac{dS}{(PQ)^2} = d\Omega \ ,$$

the result is

$$\frac{m\,a^2\,d\Omega}{(OP')^2} .$$

The sum of such contributions from the entire shell is simply

$$\frac{4\pi\,m\,a^2}{(OP')^2} = \frac{M}{(OP')^2} \ ,$$

where M is the total mass of the shell.   Q.E.D.

A solid spherical body is a composite of subshells. Hence, the attractive force on a unit mass at any external point P' is directed toward the center of the spherically symmetric body (with any radial dependence of its density) and has the same magnitude as though the entire mass of the body were located at its center—a result of fundamental importance in celestial mechanics.

66. (a) Has been determined by a laboratory experiment.

67. (d) Proportionality constant in Newton's law of gravitation.

68. (a) Directed toward the Sun.

69. (a) Newton's law of gravitation and his three laws of motion are believed to be valid throughout the universe and hence equally applicable to the movement of planets around the Sun and of satellites around a planet.
    (b) The mass of the central body is appropriately different.

70. (c) Always directed toward the center of the Earth.

71.

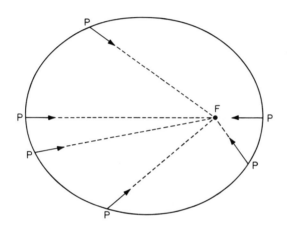

An elliptical planetary orbit about the occupied focus F is illustrated. The vectors show the directions, but not the magnitudes, of the accelerations at various points P along the orbit. Inasmuch as the attractive forces lie along the respective lines PF, so also do the acceleration vectors. The magnitudes of the accelerations are inversely proportional to the squares of the respective radial distances PF. The acceleration picture is the same for either sense of the planet's orbital motion, clockwise or counterclockwise.

72. The complete roster of conic sections is the following: point, straight line, circle, hyperbola, parabola, and ellipse. The spirit of this question favors hyperbola and parabola as answers.

73. All three types of orbits in a central gravitational field are conic sections with varying shapes as specified by their periapsidal distances and eccentricities. The planes of their orbits pass through the center of the force field.

■ An elliptical orbit is a closed one, characterized by negative total energy (sum of kinetic energy and gravitational potential energy) and eccentricity between zero and 1.0. A circular orbit is a special case of an elliptical one with eccentricity zero.

■ A parabolic orbit is an open one, characterized by zero total energy and eccentricity 1.0.

■ A hyperbolic orbit is also an open one, characterized by positive total energy and eccentricity greater than 1.0.

74. Perihelion.

75. Newton's law of gravitation.

76. Irrespective of the velocity vector of any one of these bodies, the gravitational attraction of the Sun is a vector force lying along the line from the body to the Sun. By Newton's second law of motion, the resulting acceleration vector lies along the same line. Hence the change in the velocity vector lies in the plane defined by the velocity vector and the force vector; and the velocity vector continues to lie in the same plane. This plane passes through the Sun and remains fixed in space as the motion proceeds.

77. In the suggested halo orbit, the acceleration vector would lie in the plane of the orbit. But this is physically impossible because the attractive gravitational force of the Earth and the consequent acceleration vector pass through the center of the Earth at every moment.

78. This is only a rough approximation to an inertial coordinate system primarily because the origin is continually accelerated toward the center of the Earth and also because the Earth is continually accelerated toward the Sun. An astronomical coordinate system whose origin is at the barycenter of the solar system and whose axes point to distant stars is the best available approximation to a true inertial coordinate system.

79. (d) Is essential in understanding the motion of the Moon.

80. (b) The period of revolution of a planet is approximately independent of its mass; and
    (d) All objects in the solar system affect all other objects.

81. The gravitational force per unit mass is the same for the astronaut as for the spacecraft. Hence their acceleration vectors and their initial velocities are identical and the two objects continue to traverse identical orbits at the same rate (ideally). The same considerations account for the fact that an astronaut or a piece of loose equipment interior to a spacecraft appears to be "weightless" (actually in a state of free fall as is the spacecraft itself).

82. The acceleration vector is directed to the center of the circular orbit. Its magnitude is given by

$$a = \frac{v^2}{r}$$

83. For a given orbital radius, the orbital speed is calculable from first principles and is the same for any orbiting object. Hence, its value is not news.

84. The magnitude of gravitational acceleration toward the Sun is the ratio of the gravitational force to the mass m of the planet,

$$a = \frac{F}{m} = \frac{GM}{r^2} \, , \tag{1}$$

where G is the universal gravitational constant and M is the mass of the Sun. Also, as given by the kinematic equation,

$$a = \frac{v^2}{r} \, . \tag{2}$$

By (1) and (2)

$$v^2 = \frac{GM}{r} \quad \text{or} \quad v \propto \frac{1}{\sqrt{r}} \, . \tag{3}$$

85. Newton's laws of motion and the fact that the Sun's mass is far greater than the Earth's make it obvious that, in the relative motion of the two bodies, the Earth's motion is much greater than is the Sun's, i.e., the Sun is effectively stationary.

    Direct confirmation of the Earth's orbital motion about the Sun is provided by the classical observations of the annual variation of the aberration of starlight and of the parallax of relatively nearby stars and by the annual variation of the Doppler shift of lines in stellar spectra.

    Modern techniques in the flight dynamics of artificially launched spacecraft and the Doppler shift of radio signals therefrom provide massive and precise confirmation of the orbital motion of the Earth.

86. (c) A small object placed at X will initially move away from X at the velocity it has relative to an inertial coordinate system.

87. The object is subjected to the gravitational force of the Sun and, usually to a lesser extent, to the gravitational forces of all other objects in the solar system. Hence, it is accelerated in the direction of the resultant force and acquires motion even if initially at rest.

88. It is commonly stated that the mean distance from the Sun to a planet is equal to the semimajor axis a of the planet's elliptical orbit about the Sun. This statement does not specify the nature of the mean and is therefore enigmatic. The diagram below aids in visualizing the enigma.

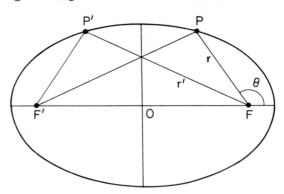

The Sun is at the occupied focus F of the ellipse and P is an arbitrary position of the planet at radial distance r = FP and angle θ from the major axis. The magnitude of r is a function of θ or a function of time t from perihelion passage *or* a function of arc length s along the ellipse from perihelion. Thus, there are at least three plausible bases for computing a mean value of r.

The three corresponding mean values of r are:

$$\langle r \rangle_\theta = \frac{\int r(\theta)d\theta}{\int d\theta} \, , \tag{1}$$

with the integrals taken over the range from $\theta = 0$ to $\theta = 2\pi$;

$$\langle r \rangle_t = \frac{\int r(t)dt}{\int dt} \, , \tag{2}$$

with the integrals taken over the range from t = 0 to t = P, the orbital period; and

$$\langle r \rangle_s = \frac{\int r(s)ds}{\int ds} \, , \tag{3}$$

with the integrals taken over the range of arc length s along the ellipse from s = 0 to s = circumference of the ellipse.

Of these three, only equation (3) yields the result that

$$\langle r \rangle = a \, . \tag{4}$$

A simple proof of (4) is as follows:

Imagine an infinitesimal arc of the ellipse ds at P and an equal arc ds at another point P' located on the ellipse such that P and P' are equally distant from the minor axis; and let F' be the unoccupied focus. By the basic property of an ellipse

$$FP + F'P = 2 \, a \, ,$$

and by symmetry

$$F'P = FP' \, .$$

Hence,

$$FP + FP' = 2 \, a \, ,$$

so that the average value of the two radial distances r and r' is

$$\left(\frac{FP + FP'}{2}\right) = a .$$

Inasmuch as this is a general result for an arbitrarily chosen point P, it can be generalized to apply to the arc length average of r over the entire ellipse, by pairing equal arc length elements around the circumference.   Q.E.D.

89. (d) Mean distance from the Sun to the Earth.

90. ▪ The Keplerian/Newtonian theory of planetary motion together with geometric triangulation provides accurate knowledge of the geometric properties of planetary orbits in terms of the semimajor axis of the Earth's orbit, the "astronomical unit". The next essential problem is to determine the magnitude of the astronomical unit in terms of terrestrial units of length, for example, the kilometer.

▪ The classical method of doing so is a geometric one. The parallax of the Sun or, better, that of a planet or closely approaching asteroid as observed on the star field simultaneously from different points on the Earth or from the same point as the Earth rotates calibrates the astronomical unit in terms of a known terrestrial distance.

▪ Interpretation of the perturbation of the orbit of an asteroid in its close passage by a planet also determines the AU in terms of terrestrial units of distance.

▪ The measured aberration of starlight determines the ratio of the orbital speed of the Earth to the known speed of light. The orbital speed is known in AU yr^{-1}. Hence the AU is calibrated in kilometers.

▪ Precise modern methods of calibrating the AU depend on the reflection of radar pulses by planets and asteroids and on the flight dynamics of spacecraft.

91. v = 42.12 km s^{-1} (heliocentric).

92. By problem 84, the Newtonian theory predicts that $v \propto 1/\sqrt{r}$ or that $v\sqrt{r}$ is a constant. The given data correspond to the following values of $v\sqrt{r}$

	$v\sqrt{r}$
Mercury	30.0
Earth	30.0
Jupiter	29.6

93. By Kepler's third law, one expects P^2/a^3 to be the same for each of these satellites of Jupiter. This expectation is tested by the following table:

	P^2/a^3
Io	42.3
Europa	41.9
Ganymede	41.7
Callisto	41.9

94. In the circular orbit approximation, the acceleration vector is directed toward the Sun and is orthogonal at every moment to the velocity vector which is tangent to the orbit. Hence, the direction of the velocity vector is changed continuously but its magnitude remains constant. Note that a circular orbit is not the general case.

95. 120,000 years.

96. 64,500 years.

97. 7,510 minutes (125 hours).

98. (a) 21,250 minutes (14.76 days).
    (b) 7/1.

99. (a) 11.0 Earth radii.
    (b) 19/1.

100. (b) 9.9 days.

101. (d) 6 Earth radii, an elementary estimate.

102. 23.7 hours.

103. (a) 1/12.

104. (b) Atmospheric drag.

105. Air friction causes a loss of energy of the satellite and results in a more negative value of its total energy (kinetic energy plus gravitational potential energy) corresponding to a lower altitude orbit. But v $\propto$ 1/$\sqrt{r}$ so that the speed (and hence the kinetic energy) of the satellite in the orbit of radius r becomes greater as r decreases.

    This paradoxical result—namely that a loss of energy to air drag results in an increase in the kinetic energy of the satellite—finds its explanation in the dynamical equation of motion and the fact that the increase in kinetic energy is supplied by a decrease in gravitational potential energy.

106. 84.5 minutes with an orbital radius equal to the radius of the solid Earth. Because of the atmosphere practical orbits have periods of about 90 minutes or greater.

107. 6.61 Earth radii.

108. (c) 16 percent.

109. (d) Follow a slightly eccentric orbit around the Sun similar to that of the Earth.

110. 4 g.

111. (c) The Moon's orbit would be unaffected (to first order).

112. (a) Calculate the average distance from the Earth to the Moon.

113. (b) The orbits of planets would remain the same.

114. The Earth's nearly circular orbit would become an ellipse with aphelion at 1.0 AU, perihelion at 0.33 AU, eccentricity 0.50, semimajor axis 0.67 AU, and period of revolution 0.38 year.

115. The Earth's orbit would be only slightly changed because of the slight change in the center of mass of the Earth-Sun system.

116. (b) Are identical.

117. (b) Produce effects that have been confirmed by observation.

118. 326 Jovian radii. Note that Sinope, the outermost known satellite of Jupiter, is in a retrograde orbit of semimajor axis 332 Jovian radii and eccentricity 0.28, thus posing a challenging problem in celestial mechanics.

119. (d) The rings of Saturn consist of several thin, rigid sheets of matter revolving at different rates around the axis of the planet. (False)

120. (a) Go into a quite elliptical orbit, because the injection speed will be greater than that necessary for a circular orbit.

121. If the engineer neglects to take account of the rotation of the Earth, the calculated value of the launch (injection) speed will exceed that necessary for a circular orbit. Hence, the satellite would go into an elliptical orbit with perigee at the injection site.

122. Consider the case in which, at the launch (injection) point, the velocity vector of the satellite is directed due East. The resulting orbit is in the plane containing the velocity vector and the gravitational force vector through the center of the Earth. This plane is inclined at 28.3 degrees to the equator. If the injection velocity vector lies *either* North *or* South of due East, the plane of the orbit is inclined at an angle greater than 28.3 degrees.

123. Following delivery of the satellite into a low altitude circular orbit inclined at, say, 28.3 degrees to the equator, the following additional steps are necessary:

■ The satellite is propelled into an elliptical orbit with perigee at low altitude and apogee at 6.6 Earth radii. This is called the geosynchronous transfer orbit.

■ The resulting orbit must be circularized at about 6.6 Earth radii.

■ The original, inclined plane of the orbit must be changed to the plane of the equator. This step must be accomplished on the equator.

By detailed consideration of the various ways in which the second and third steps may be taken, one finds that it is most advantageous to accomplish both at apogee of the transfer orbit inasmuch as the maneuvers there require lesser impulses than at any other point. Thus, the first step should be such that the perigee and hence the apogee of the transfer orbit are on the equator.

124. 0.177 year.

125. (d) In the direction opposite to the Earth's orbital motion at a net escape speed of 30 km s^{-1}.

126. (b) An ellipse.

127. The Pioneer 10 mission (1972−present) provides an important and informative example of the principle of gravitational assist. The following discussion of that mission, though simplified in detail, is of adequate accuracy to faithfully describe the principle.

■ Pioneer 10 was launched on 3 March 1972 with a heliocentric speed of 39 km s^{-1} (after escape from the Earth's gravitational field) in the direction of the Earth's motion. The resulting heliocentric orbit was an elliptical one having the following parameters:

Perihelion distance	1.0 AU
Aphelion distance	6.0 AU
Semimajor axis	3.5 AU
Eccentricity	0.7145
Period	6.55 years

■ The spacecraft made the first ever encounter with Jupiter on 4 December 1973, 21 months later. The encounter was a prograde flyby, the closest radial distance of approach being 2.84 planetary radii. In the heliocentric sidereal coordinate system, the spacecraft had a velocity of 9.8 km s^{-1} (before it entered Jupiter's sphere of gravitational influence) at an angle of 49 degrees East of the outward drawn radial vector from the Sun. At the time of the encounter the radial distance of Jupiter was 5.05 AU and its speed was 13.5 km s^{-1}. Jupiter crossed the orbit of the spacecraft slightly before the latter's arrival at that distance.

■ The spacecraft's asymptotic approach vector was such as to yield a hyperbolic orbit in a sidereal coordinate system centered on the planet. The magnitude and direction of this vector are derived with the help of the following vector diagram:

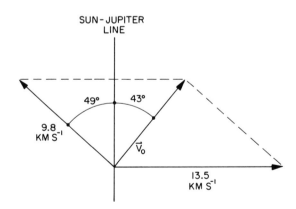

In the diagram, $\vec{V}_0$ is the asymptotic approach vector in the planetocentric coordinate system. $|V_0| = 8.9$ km s^{-1} at an angle of 43 degrees West of the Sun-Jupiter line as shown.

■ In the planetocentric coordinate system the asymptotic velocity vector $\vec{V}_1$ after encounter and emergence from Jupiter's gravitational sphere of influence had the same magnitude as did $\vec{V}_0$ but was rotated clockwise by 116 degrees.

■ The resulting velocity vector $\vec{V}'$ of the spacecraft in the heliocentric coordinate system is derived from the following vector diagram:

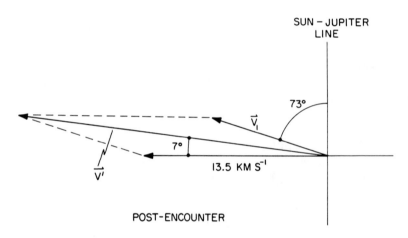

SUN – JUPITER LINE

73°

$\vec{V}_1$

7°

13.5 KM S^{-1}

$\vec{V}'$

POST-ENCOUNTER

■ $\vec{V}'$ had a magnitude of 23 km s^{-1} and was directed at an angle of 83 degrees East of the Sun-Jupiter line. Thus the encounter resulted in an increase of heliocentric speed from 9.8 to 23 km s^{-1} and a marked change in direction. The dynamics of the encounter show a corresponding, but undetectably small, loss of the planet's orbital energy.

■ The heliocentric escape speed at 5.05 AU is 20 km s^{-1}. Hence it is clear that the post-encounter orbit of Pioneer 10 was a hyperbolic one with respect to the Sun and that the spacecraft was destined to escape the solar system at an asymptotic rate of 2.39 AU yr^{-1} (radial speed at infinite distance). The progress of such an escape has been observed continuously during subsequent years, through 1992, and can be projected with confidence thereafter.

■ As illustrated by this example, the principle of gravitational assist has many practical applications in the flight of spacecraft as well as in understanding natural phenomena.

128. 2.60 years.

129. 10.27 years.

130. (d) 84 years.

131. 0.70 year.

132. 31.6 years.

133. (c) 4.9 days.

134. 96 days later.

135. 5.70 AU.

136. 248 years.

137. 1.42 years.

138. 29.3 years.

139. 39.5 AU.

140. 17.97 AU.

141. Bode's law is an empirical one that gives an approximate relationship among the semimajor axes of planetary orbits. It has no fundamental physical foundation, though it may be suggestive of one. The explicit statement of Bode's law is

$$a_i = \frac{k + 4}{10}$$

wherein k = 0, 3, 6, 12, 24, etc. for successive planets Mercury, Venus, Earth, Mars, etc. and a_i is the semimajor axis in AU of the orbit of the ith planet numbered outward from the Sun. Note that the asteroid belt, but no major planet, corresponds to i = 5 and k = 24.

142. Mercury, Venus, Earth, Mars, Jupiter, Saturn, Uranus, Neptune, Pluto.

143. 165 years.

144. Radius of orbit = $(1047)^{-1/3}$AU = 1.473 × 10^7 km
             = 206 Jovian radii.
    Would such a hypothetical satellite be stolen away by the Sun? (cf. problem 118 and tables of known satellites of Jupiter)

145. 236 days.

146. 6.83 times the radius of the Moon's orbit
          = 2.62 × 10^6 km
          = 36.7 Jovian radii.

147. 1.79 hours.

148.

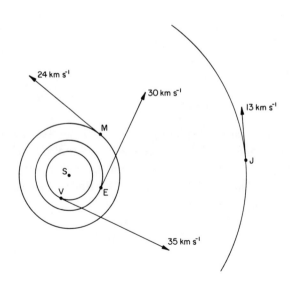

This approximately-to-scale drawing shows the orbits about the Sun of Venus, the Earth, Mars, and Jupiter and the velocity vectors of these planets at a particular moment. Hence the vectors represent the speeds and directions of the respective motions of the planets if the Sun suddenly ceased to exist.

149. The Earth would move along a parabolic escape orbit (eccentricity = 1.0) out of the solar system with perihelion at 1.0 AU.

150. The orbital speed of the Earth is 2π AU yr^{-1} and of Neptune
$$\frac{(2\pi)(30)}{(164.8)} \text{ AU } yr^{-1} .$$
Each planet would move along one side of a right triangle whose base is 1.0 AU for the Earth and 30 AU for Neptune and whose hypotenuse is heliocentric distance. The respective times to reach a heliocentric distance of 50 AU would be 7.96 years for the Earth and 35.0 years for Neptune.

151. 7.96 years.

152. Kepler's third law yields the familiar relationship between orbital period P, radius of orbit a (equals radius of the planet in this case), mass of the planet M, and the universal gravitational constant G:
$$P^2 = \frac{4\pi^2}{GM} a^3 . \qquad (1)$$
Also,
$$M = \frac{4\pi a^3 \bar{\rho}}{3} \qquad (2)$$
where $\bar{\rho}$ is the mean density of the planet, the ratio of its total mass to its total volume.

Combining (1) and (2), one finds that
$$P^2 = \frac{3\pi}{G\bar{\rho}} . \qquad (3)$$
Hence,
$$P \propto \frac{1}{\sqrt{\bar{\rho}}} , \qquad (4)$$
independent of a. $\qquad\qquad$ Q.E.D.

153. As shown in the answer to problem 152, the period of revolution of a satellite in a close orbit varies as the inverse square root of the planet's mean density $\bar{\rho}$ and is independent of the radius of the planet. Among the major planets, $\bar{\rho}$ is the greatest $(5.518 \text{ g cm}^{-3})$ for the Earth and the least (0.70 g cm^{-3}) for Saturn. The ratio of the inverse square roots is 0.356. Thus P = 1.4 hr. for the Earth and 3.9 hr. for Saturn. The values of P for all other planets (and the Moon) lie within the range 1.4 to 3.9 hr.

154. 71 minutes.

155. 3.75 Martian radii.

156. At any distance r from the center of the body, the gravitational force F on the object is directed toward the center and is the same as though the mass of the body interior to radius r were concentrated at the center (cf. problem 65). Thus, with usual symbols,

$$F = -\frac{Gm}{r^2}\frac{4\pi r^3 \rho}{3} \tag{1}$$

or

$$F = -\frac{4\pi \rho Gm}{3} r \ . \tag{2}$$

Equation (2) is recognized as being the equation of a simple harmonic oscillator whose period

$$P = 2\pi \sqrt{(3/4\pi \rho G)} \ . \tag{3}$$

The total mass of the body

$$M = \frac{4\pi a^3 \rho}{3} \tag{4}$$

so that (3) can be rewritten as

$$P^2 = \frac{4\pi^2 a^3}{GM} \ . \tag{5}$$

It is noted that the period of oscillation given by (5) is identical to the period of revolution of a satellite of the body in an orbit of radius a.

157.

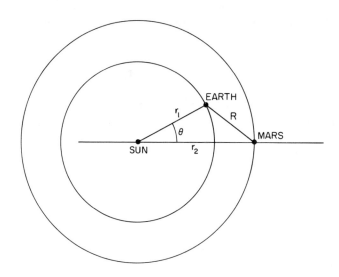

Denote the radius of the Earth's orbit by r_1, the radius of Mars' orbit by r_2, and their respective orbital speeds by v_1 and v_2. Adopt the opposition of Mars as the starting condition and denote the subsequent heliocentric angle between the Sun-Earth line and the Sun-Mars line by θ. The time rate of change of θ is constant and is given by

$$\frac{d\theta}{dt} = \frac{v_1}{r_1} - \frac{v_2}{r_2} . \tag{1}$$

The distance between the two planets D is given by

$$R^2 = r_1{}^2 + r_2{}^2 - 2r_1 r_2 \cos \theta . \tag{2}$$

The maximum value of $dR/d\theta$ occurs when the Mars-Earth line is tangent to the Earth's orbit. Its value at that point is simply

$$\frac{dR}{d\theta} = r_1 . \tag{3}$$

The time rate of change of the distance between the two planets is given, in general, by

$$\frac{dR}{dt} = \frac{dR}{d\theta} \frac{d\theta}{dt} \tag{4}$$

and therefore, by (1) and (3), its maximum value is

$$v_1 - v_2 \frac{r_1}{r_2} . \tag{5}$$

The numerical value of this maximum is 14.2 km s^{-1}. At the moment of maximum dR/dt, θ = 48.9 degrees, and R = 1.145 AU.

158. Ring plane crossings occur during periods of the order of nine-months' duration, separated by intervals of approximately one-half the orbital period of Saturn, i.e., at average intervals of 14.7 years. During each period, at least one and at most three crossings occur. Two crossings represent a case of zero measure.

On the basis of the given data, at least one and perhaps as many as three crossings would be expected in early 1981 or thereabouts.

The next set of actual crossings occurred on 27 October 1979, 13 March 1980, and 23 July 1980, about a year earlier than average because of the eccentricity of Saturn's orbit with perihelion passage in early 1974.

159. Saturn's ring plane passes through the Sun once during each of the planet's half-orbits. The average lapse of time between such ring plane crossings is one-half of the planet's orbital period, i.e., 14.7 years. Hence, the average expectation would be that the next occasion would have been in early 1981.

The next actual occasion was on 2 March 1980, about a year earlier than average because of the eccentricity of Saturn's orbit with perihelion passage in early 1974.

160. ■ The plane of the rings of Saturn is tilted at an angle of 26.7 degrees to the plane of its orbit. The intersection of these two planes defines a line of nodes. As Saturn moves along its orbit it carries this line of nodes with it, the line remaining parallel to itself. Whenever Neptune crosses the line of nodes, it lies in the plane of Saturn's rings.

■ Let the common plane of the specified orbits of Saturn and Neptune be the X-Y plane and let the line of nodes be parallel to the X-axis. For Saturn,

$$y_1 = 10 \sin 12\, t \tag{1}$$

and for Neptune,

$$y_2 = 30 \sin 2.1818\,(t - t_0) \tag{2}$$

where t_0 may be chosen to provide different phasing between the two orbits and t and t_0 are in years. A hypothetical observer on Neptune can see Saturn's rings edge-on whenever

$$y_1 = y_2 . \tag{3}$$

■ The corresponding transcendental equation can, of course, be solved for t numerically. A sample graphical treatment of the problem is shown in the diagram below, for $t_0 = 30$ years. If the two sinusoids are plotted on separate sheets and slid over each other on a light table, the effect of varying t_0 can be examined. It is found that at least two edge-on apparitions of Saturn's ring plane occur during each orbit of Neptune.

But as shown in the diagram as many as four are possible, one bundle of three and a single crossing about 165/2 years later. A slightly different set of parameters results in two bundles of three or a total of six apparitions during an orbital period of Neptune.

■ A similar diagram is applicable to the Earth-Saturn system.

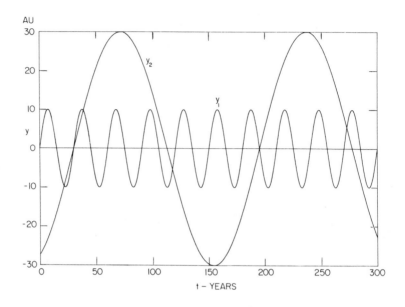

161. 39.44 AU.

162. The greatest value of the eastern or western elongation of Mercury (angle between Sun and Mercury as seen from the Earth) occurs when Mercury is at its aphelion, the Earth is at its perihelion, and the Sun-Mercury-Earth angle is 90 degrees. The value is

$$\arcsin\left(\frac{0.4667}{0.9833}\right) = 28.3 \text{ degrees} .$$

163. (c) Venus.

164.

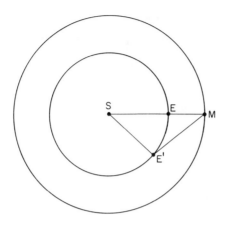

Suppose that Mars' synodic period of revolution (e.g., the lapse of time between successive oppositions) has been observed and that its corresponding sidereal period has been calculated. An opposition is represented by the straight-line configuration SEM in the diagram. After a lapse on one Martian sidereal period, Mars will have returned to the same point of its heliocentric orbit but the Earth will be at a different point E', with a known angle ESE'. At that time the geocentric angle SE'M can be measured. In the triangle SE'M, there is the following relationship of angles:

$$E'MS = 180 - (ESE' + SE'M) , \qquad (1)$$

and the sine formula yields

$$\frac{SM}{(\sin SE'M)} = \frac{1.0}{(\sin E'MS)} \qquad (2)$$

because SE' = 1.0 AU. By means of (2), SM, the radius of Mars' orbit, can be calculated in terms of astronomical units. Alternatively SM can be measured on a carefully laid out drawing.

The above is a somewhat simplified version of the actual situation but properly represents the essence of the method. Knowledge of the distance SM in terrestrial units, e.g., kilometers, must await calibration of the AU (cf. problems 90 and 167).

165. Kepler's method of finding the radius of Mars' orbit is essentially that of the answer to problem 164 (q.v.)

166. 18,500.

167. By determining the round trip lapse of time between transmission of a radar pulse and reception of the reflected pulse from a planet when the planet is in a known configuration with respect to the Sun and the Earth such that planet-Earth distance is a known number of AU. The speed of light is assumed to be known.

168. (c) Eastward by about 12 degrees per year.

169.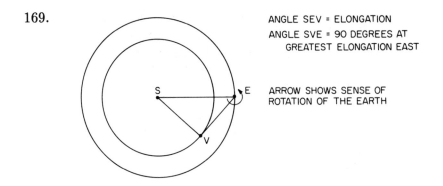

170. (c) It is at eastern elongation.

171. (a) Mercury.

172. (b) Inferior conjunction.

173. 0.73 AU.

174. A personal observation.

175.

176. Opposition.

177. (d) Opposition.

178. (a) Opposition.

179. (b) Venus.

180. (a) Venus.

181. See answer to problem 42.

182. (c) Opposition.

183. (a) Inferior conjunction.

184. Inferior conjunction.

185. (c) Opposition.

186. The Sun was near the line of sight from Mars to the Earth so that the telemetered signals from the spacecraft at Mars were obscured by solar radio emissions.

187. (b) Venus.

188. (b) Inferior conjunction.

189. 0.47 AU.

190. (b) Its *average* motion is eastward.

191.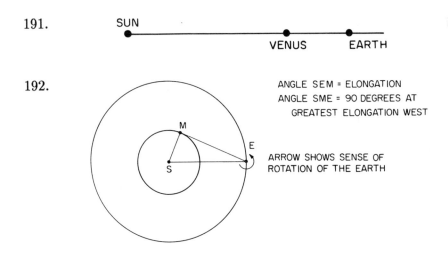

192.

193. The lapse of time between successive identical or nearly identical configurations of the Sun-Earth-planet system, e.g., between successive superior conjunctions of an inferior planet or successive oppositions of a superior planet.

194. (b) 4.7 months.

195. 586 days.

196. 578 days (twice the interval from superior conjunction to inferior conjunction).

197. 584 days.

198. 116 days.

199. 687 days.

200. 88 days.

201. 399 days.

202. 2.0 years.

203. 0.653 year.

04. 584 days (same as the time between successive inferior conjunctions of Venus as viewed from the Earth).

05. (c) 335 days.

06. 120, 115, and 119 days, averaging 118 days.

07. 109, 117, 129, and 120 days, averaging 119 days.

08. 120, 116, 116, 126, 107, 120, 115, 105, and 113 days, averaging 115 days.
    [Note: Problems 206, 207, and 208 show the variability of the synodic period of Mercury, caused principally by the large eccentricity (0.2056) of its orbit. The long-term mean value of its synodic period is 115.9 days.]

09. 0.50 year.

10. In this case the sidereal period of revolution equals the synodic period = 2.0 years. According to the geometric specification of the problem, the hypothetical planet has an orbit of radius 2.0 AU. But, by Kepler's third law, a sidereal period of 2.0 years corresponds to an orbital radius of 1.59 AU. Or conversely, an orbital radius of 2.0 AU corresponds to a sidereal period of 2.83 years. Hence, the hypothetical planetary orbit is physically impossible.

11. 72.2 days.

12. 20 years.

13. 179 years.

☆ ☆ ☆

ANSWERS TO

CHAPTER 3

Sun, Earth, and Moon

214. G, the universal constant of proportionality in Newton's law of gravitation.

215. By measuring the attractive force between two known masses m_1 and m_2 separated by a distance r and using Newton's law of gravitation

$$F = G \frac{m_1 m_2}{r^2} ,$$

one can find the value of G.

Then one can measure the acceleration g of a freely falling body in vacuo at the surface of the Earth.

Finally, the relationship $g = GM/r^2$, where M is the mass of the Earth and r its radius, determines M in terms of known values of g, r, and G.

216. 4,000 miles.

217. An idealized method is to make simultaneous measurements of the meridian transit altitude of a star, the Sun, or a planet at two stations at the same longitude but separated along the surface of the Earth by a known North-South distance s. If the difference of the two observed altitudes is Δh radians, then the radius of the Earth is given by $r = s/\Delta h$, as can be confirmed by a simple diagram. There are obvious variations of the above idealized scheme to yield the same result.

218. 1.852 km (or 1.000 nautical mile).

219. 19.6 km.
     The navigator's convenient version of the given formula is $d = 1.1\sqrt{H}$ where d is in nautical miles and H in feet.

220. (d) 210 miles.

221. (a) 1,700 ft.

222. 1,304,000.

223. (a) Study of the relative abundances of various radiogenic elements in granites.

224. (b) 4 billion.

225. (a) 4×10^9 years.

226. (b) 4.5 billion years ago.

227. 3.9×10^9 years.

228. 4.5×10^9 years.

229. 0.25 g.

230. The Earth is composed mostly of rocky material (and some iron) under the high pressure of its self-gravitation.

231. Its mean density 5.52 g cm^{-3} is substantially less than even the uncompressed density of iron, namely 7.8 g cm^{-3}, and much less than its compressed value.

232. (b) Its rotation.

233. 68 degrees Kelvin.

234. The heat released by the decay of radioactive substances and the residual (gravitational) heat from its original formation by the accretion of dispersed material.

235. Volcanic eruptions of molten lava, hot water from geysers, and measured temperatures in deep oil wells.

236. Because of variable refraction along the observer's line of sight through the atmosphere.

237. (a) Scattering.

238. (a) Sunlight scattered in the atmosphere.

239. (c) Refraction.

240. (d) The apparent altitude of a star to be greater than its true (geometric) altitude.

241. (c) 3.0.

242. 11.9 meters. The density of liquid air is about 0.87 g cm^{-3}.

243. (d) Nitrogen.

244. (a) Ozone is O_3 or triatomic oxygen.
(b) 25 km.
(c) It absorbs actinic ultraviolet light from sunlight, which would otherwise cause severe sunburn and possibly cancer on exposed human skin; cataracts in eyes; and other adverse effects on biological material.

245. The minimum orbital period of any free-flying satellite of the Earth is 84.5 minutes, in an orbit of radius equal to the radius of the Earth (neglecting atmospheric drag). Any higher altitude orbit has a greater period. In principle, a continuously propelled aircraft with "upside down" wings to cause a negative aerodynamic lift could fly a great circle path around the Earth in the upper atmosphere in less than 84.5 minutes.

246. 490 meters (1,600 feet, nearly three times the height of the Washington Monument).

247. (c) Electrical currents in its interior.

248. (b) The solar wind.

249. (a) Would not exist if the Earth were not magnetized.

250. A durably trapped electrically charged particle travels at constant energy in a spiral path around a magnetic line of force, with a typical period of the order of a millisecond, and oscillates back and forth between "mirror points" in opposite hemispheres with a typical period of a second. The particle's spiral path drifts around the planet in longitude with a typical period of an hour. The foregoing exemplary periods are for electrons in the Earth's magnetosphere. They are different for protons and other ions and for other planets.

251. Perpendicular to the plane of the Earth's orbit, the ecliptic plane.

252. The sketch illustrates the illumination of a representative portion of the Moon's rough surface at first quarter phase. It is evident that a substantial fraction of the surface is in shadow and that, in turn, only a fraction of the illuminated surface is visible from the Earth.

In contrast, at full phase a much greater fraction of a representative portion of the Moon's surface is illuminated and nearly all of the illuminated surface is visible from the Earth.

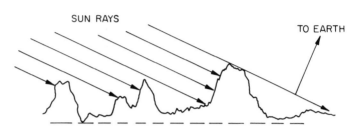

253. See answer to problem 252.

254. (b) Rises about 50 minutes later each night.

255. (d) Is full when it crosses the meridian at midnight local time.

256. (d) Perigee.

257. About 28 days.

258. (d) Ecliptic.

259. (c) Perigee.

260. (b) Eastward along the ecliptic at about 13 degrees day^{-1}.

261. ■ As viewed in the rotating coordinate system attached to the center of mass of the object and with one axis through the center of the Earth, the orbiting object as a whole may be described as in a state of dynamical equilibrium between the inward gravitational force $2GMm/r^2$ and the outward centrifugal force $2m\,r\,\omega^2$ so that

$$\frac{2GMm}{r^2} = 2m\,r\,\omega^2 \, , \tag{1}$$

where r is the radius of the orbit; G the gravitational constant; M the mass of the Earth; ω the angular orbital speed of the object; and 2m the mass of the object. Equation (1) reduces to the familiar Keplerian/Newtonian relationship.

$$\omega^2 = \frac{GM}{r^3} \tag{2}$$

as derived in an inertial coordinate system.

■ Continuing the analysis in the rotating coordinate system and supposing that the rod makes a small angle θ to the radial line, one finds that there is a net downward force on the lower mass equal to

$$\frac{GMm}{(r - \ell/2)^2} - m(r - \ell/2)\omega^2 \tag{3}$$

and a net upward force on the upper mass equal to

$$-\frac{GMm}{(r + \ell/2)^2} + m(r + \ell/2)\omega^2 \ . \tag{4}$$

For $\ell/2 \ll r$, expression (3) reduces to

$$\frac{3m\omega^2\ell}{2} \ , \tag{5}$$

and expression (4) also reduces to

$$\frac{3m\omega^2\ell}{2} \ . \tag{6}$$

These two equal and opposite forces with a lever arm $\ell \sin \theta$ ($\simeq \ell\theta$) constitute a couple whose torque T on the object is

$$T = -\left(\frac{3m\omega^2\ell^2}{2}\right)\theta \tag{7}$$

and whose sense is such as to reduce θ.

- The foregoing analysis was restricted to small values of θ but it is evident that even for large values of θ, the sense of the torque is to reduce θ (except for the singular case $\theta = \pi/2$). Hence the equilibrium orientation of the object is with the rod lying along the radial line through the center of the Earth. Q.E.D.

- The moment of inertia of the object about its center of mass is

$$\frac{m\,\ell^2}{2}\,, \tag{8}$$

and (7) is recognized as the equation of a simple harmonic oscillator whose period is

$$p = \frac{\left(\dfrac{2\pi}{\omega}\right)}{\sqrt{3}} \tag{9}$$

or

$$p = \frac{P}{\sqrt{3}} \tag{10}$$

where P is the orbital period of the object. For example, if P = 90 minutes, p = 52 minutes. The tension in the rod in the equilibrium configuration is

$$\frac{3m\,\omega^2\ell}{2}\,. \tag{11}$$

- The so-called gravity-gradient (passive) control of the orientation of satellites is illustrated by the above simple example. The analysis is also applicable to the case of a tethered satellite—a system consisting of a large orbiting body and a smaller one connected by a tether. In some proposed applications the length of the tether is planned to be as great as 20 km.

- The orientation of the Moon with one face toward the Earth is a kindred tidal torque phenomenon because of the slightly asymmetric distribution of the Moon's mass.

262. (c) Synodic month.

263. 29.53 days.

264. (a) Is rotating counterclockwise (as viewed from the northern celestial pole) with a period of 27.32 days.

265. (a) Its sidereal period of revolution.

266. (a) Full.

267. (d) Moves mainly toward the East.

268. By observing and plotting the positions of the Moon on the star field over a period of a few weeks and finding the interval of time between its having the same position.

269. Observe the positions of the Moon on the star field over a period of a few weeks and plot these positions on a star chart. Measure the maximum angles North and South of the ecliptic. Each is equal to the inclination of the Moon's orbit to the ecliptic.

270. (c) Its right ascension differs by the same amount from the right ascension of the Sun.

271. Of the order of one-fourth. The observer's hemisphere subtends a solid angle 2π steradians, whereas the full Moon subtends a solid angle of 1/15,600 steradian. Hence 98,000 full Moons would just fill the observer's hemisphere. The exact ratio of the illuminations on a horizontal surface depends on the altitude of the Sun and on a proper average of the obliquity of the light from the array of Moons.

272. (a) 115.

273. Late December–early January.

274. (c) Sometimes occurs near the descending node.

275. July (plus or minus two or three months, but seasonal span varies from year to year).

276. (a) Winter solstice.

277. The Moon, irrespective of phase, crosses an observer's median at an average interval of 24 hours 50 minutes. Hence, there are about $365.26/1.0347 = 353$ lunar days in a solar year.

278. (a) 121 days.

279. Tides on the Earth, monthly wavering of the apparent position of a nearby asteroid on the star field, optical occultation of stars, occultation of stellar radio sources, reflection of radar signals, precession of the equinox, perturbation of trajectories of planetary spacecraft.

280. Any rigid body has three mutually orthogonal principal axes of inertia. For a spherically symmetric body, the origin of the axes is at the geometric center of the body and the orientation of the system of principal axes is arbitrary. This simple situation is not characteristic of the distribution of the Moon's mass. Indeed, the Moon has three slightly different principal moments of inertia. Hence the term triaxial. Consequences of this fact are the existence of a tidal torque by the Earth and physical libration of the Moon.

281. (a) The apparent monthly wavering of the orbits of asteroids passing nearby.

282. ± 28.5 degrees maximum; ± 18.5 degrees minimum.

283. Because of the rotation of the Earth and the fact that it is not perfectly rigid, the meridian cross section of its surface is not a circle but is approximately an ellipse, with equatorial radius 6378.164 and polar radius 6356.779 km. The corresponding figure of revolution about the polar axis defines "sea level", the surface perpendicular to the vector resultant of gravitational and centrifugal forces on a parcel of fluid. This is the reference level that determines the flow of water in a river. The source of the Mississippi River is about 220 meters above sea level whereas its mouth is at sea level and thus the flow of water is indeed downhill on the rotating Earth.

284. 76.5 degrees.

285. $00^h\ 04^m$.

286. (b) 18^h.

287. (d) Midnight.

288. (d) In the western sky about an hour after sunset.

289. 06^h.

290. Full.

291. First quarter or last quarter.

292. As viewed from the Sun, the Moon's terminator is the circle bounding the illuminated hemisphere. As viewed from the Earth, the terminator is the projection of half of this circle on the plane perpendicular to the Earth-Moon line, i.e., a semi-ellipse with major axis equal to the Moon's diameter. At full moon it is a full circle.

293. (d) 06^h.

294. There is at least one plausible conjecture concerning the role of the Moon in influencing the evolution of life on the Earth. The ebb and flow of lunar tides alternately exposes and submerges the ocean beaches. It seems conceivable that this effect might have led to the adaptation of oceanic organisms to either condition and eventually to the development of amphibians.

Other, perhaps less plausible, possibilities include (a) lunar tidal influences on triggering earthquakes, eruption of volcanoes and geysers; and (b) the monthly cycle of moonlight which might have had some influence on the activity and mating habits of nocturnal animals.

295. The drawing is incorrect. Under the specified conditions, the illuminated crescent is to the lower right.

296. (d) Noon.

297. (a) Midnight.
     (b) 06^h.

298. Midnight.

299. (b) 03^h 45^m.

300. (c) In the early evening after sunset.

301. (b) Midnight.

302. Not possible if the observer is at mid-latitude. The full Moon
     rises at about the time of sunset at mid-latitudes. However, it
     is possible for the full Moon to rise at midnight in the arctic,
     though in the South not the East; or in the antarctic, though in
     the North not the East.

303. (d) Could have been correct if in the arctic.

304. (b) Never sees the Earth.

305. (c) 24^h 50^m.

306. (c) Waxing crescent.

307. (b) First quarter.

308. Last quarter and first quarter.

309. (a) The Moon is new.
     (c) The astronaut's selenographic latitude is between 7 degrees
     North and 7 degrees South.
     (d) The astronaut will not be able to see the Sun for about
     seven days.

310. Last quarter.

311. (a) First quarter.

312. Waning gibbous.

313. (a) Last quarter.

314. If the terminator, the boundary between light and dark portions of the Moon's disc, is elliptical with major axis along a diameter of the disc, the photograph shows a phase of the Moon. If, on the other hand, the terminator is a portion of a circle whose radius is about 2.7 times that of the Moon, the photograph shows a partial eclipse, i.e., the Earth's umbral shadow (cf. problem 384).

315. Some sunlight is refracted and scattered by the Earth's atmosphere so as to illuminate the Moon faintly even during a total eclipse. The intensity of such light is far less than that of direct sunlight even for a clear atmosphere; and it is further reduced by clouds, volcanic dust, etc.

316. (a) All dark.
     (b) All dark.
     (c) All dimly lighted.
     (d) All brightly lighted.

317. 29.53 days (a synodic month).

318. The sequence of phases of the Earth as viewed from the Moon— new Earth, first quarter Earth, full Earth, last quarter Earth—is identical to the sequence of phases of the Moon as viewed from the Earth except that it is shifted in time by one-half of the synodic period of revolution of the Moon.

319. During the century 1951–2050 there are 41 blue Moons, for example, in 9/1993, 7/1996, 1/1999, and 3/1999.

320. (c) Has fewer and smaller maria than does the "front" side.

321. (b) 35,000 km.

322. (c) Parallax.

323. Parallax.

324.

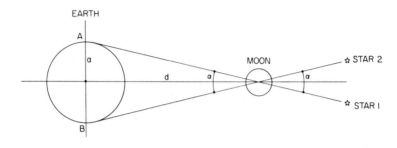

(NOT TO SCALE)

Imagine two observers, A and B, located at (almost) diametrically opposite points on the Earth and therefore separated by a distance 2a, twice the Earth's radius. Suppose that the line of sight from A to the center of the Moon pierces the star field at the position of Star 1; and simultaneously from B, Star 2. The angle between the two stars is denoted by α. Then the Earth-Moon distance d is found from the parallactic formula:

$$d = \frac{2a}{\alpha}$$

where α is in radians.

This is an idealized method, subject to modification and some complication in practice.

325. (d) 1.91 degrees.

326. 384,400 km.

327. (d) 59.

328. 176/1.

329. (c) 1.0 km s^{-1}.

330. (b) One-half that near the Earth's surface.

331. (c) 2 degrees.

332. (b) 1.02 km s^{-1}.

333. 3,350 km.

334. (c) Venus.

335. The average rotational period of the Moon is equal to its sidereal period of revolution. But because the Moon's orbit is slightly eccentric, its angular orbital speed varies cyclically during a sidereal month (Kepler's second law). Therefore, the center of its disc as seen from the Earth rocks back-and-forth or librates by a few degrees of selenographic longitude. Other contributions to libration of the Moon result from the fact that the Moon's rotational axis is not perpendicular to its orbit and that the Moon's face is viewed from slightly different angles by an observer on the rotating Earth. There is also a more subtle form of libration, called physical libration.

   An overall result is that, over time, about 59 percent of the Moon's surface can be seen from the Earth.

336. (d) 13 g.

337. By determining the ratio of the atomic abundance of a radioactive substance to that of one of its daughter products, one can calculate the lapse of time since solidification of the sample, using laboratory data on the decay scheme (cf. problems 227 and 228).

338. (c) 30.

339. (a) Altitude zero degrees; azimuth 90 degrees.
     (b) Last quarter.

340. 45 degrees.

341. (a) Occupy an underground cave whose temperature would be constant at the approximate average temperature during a month, namely +10 degrees Fahrenheit; or enclose the spacecraft within a heavily insulated structure in order to achieve essentially the same result.
    (b) Near the times of first quarter or last quarter.

342. Near first quarter or last quarter.

343. 0.166 g.

344. (d) $1/6^{th}$ of that near the Earth's surface.

345. 600 yards.

346. (c) Determining the internal temperature of the Moon.

347. The maria whose composition is that of previously molten basalt (lava).

348. (a) Meteoric impacts.

349. (c) Meteoric impacts.

350. (c) The obscuring effects of glaciation, erosion, vegetation, etc. on the Earth.

351. (d) Basalt.

352. Let H be the height of the mountain above the nearby surface and S be the length of its shadow when the Sun is at a local zenith angle z. Then S can be measured on the photograph and H can be calculated by the relation

$$H = \frac{S}{\tan z}.$$

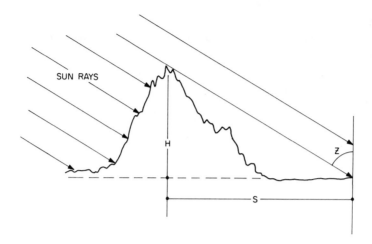

353. (b) Exclusive reliance is placed on retrorockets to reduce the landing speed.

354. (d) Retrorockets are effective despite the absence of an atmosphere.

355. (a) Similar to that of the Earth.

356. The abruptness of the occultation of a star by the Moon's disc; more recently by in situ measurements by instruments on the lunar surface.

357. The gravitational attraction on a molecule of gas at the temperature of its atmosphere is insufficient to retain it durably. If the Moon were at 10 AU from the Sun, the temperature of its atmosphere would be much less, the speed of molecular motion would be correspondingly less, and either a gaseous atmosphere or a liquid or solid coating of atmospheric constituents would be retained for long periods of time—perhaps "permanently".

358. See answer to problem 357.

359. (c) The Moon's gravitational field is too weak to prevent its escape into space.

360. (a) Xenon.
(b) Because of its greater molecular weight and correspondingly lesser speed at a given temperature.

361. (b) There is no conclusive evidence against it.

362. At the times of new and full Moon the tidal forces of the Moon and the Sun are along the same line and are additive whereas at the times of first and last quarter Moon they are at right angles to each other.

363. (a) Tidal torque.

364. Because the Moon is not a perfectly spherical body, the gravitational forces of the Earth exert a net tidal torque on the Moon. Over a sufficiently long time period energy dissipation within the body of the Moon reduces its presumed original rate of rotation to one synchronous with its orbital period.

365. (b) Near the times of both full Moon and new Moon.

366. (a) Ocean (and body) tides on the Earth.
(b) Orientation of the Moon with one face toward the Earth.
(c) Precession of the axis of rotation of the Earth (precession of the equinoxes).
(d) Slowing of the Earth's rotation.
(e) Transfer of angular momentum within the Earth-Moon system.

367. Twice this interval, namely $24^h \ 50^m$, is equal to the length of time between successive meridian transits of the Moon.

368. (a) Kepler's laws of planetary motion.

369. (a) $11^h \ 35^m$.

370. (d) $12^h \ 25^m$.

371. (c) $17^h \ 15^m$.

372. (a) Perigee passage of the new or full Moon.

373. The plane of the Moon's orbit is inclined at about 5 degrees to the ecliptic plane. The intersection of these two planes is called the line of nodes. The ascending node is the one at which the Moon moves across the ecliptic plane from South to North.

374. 0.38. The tide-raising effectiveness of a body of mass M and distance r is proportional to M/r^3, the gradient of its gravitational force on a unit mass.

375.

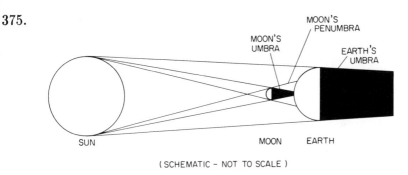

(SCHEMATIC – NOT TO SCALE)

376. Full.

377. (d) New Moon.

378. (c) Near the line of intersection of its orbital plane with the ecliptic plane.

379. An eclipse season is the interval of time during which eclipses are geometrically possible, i.e., when the intersection of the Moon's orbit with the ecliptic plane (the line of nodes) is approximately along the Earth-Sun line. Eclipse seasons are of variable duration, typically about two weeks, and recur at intervals of 173 days.

380. (b) From about one-half of the surface of the Earth.

381. (c) The orbit of the Moon were inclined at zero degrees to the ecliptic.

382. (a) Several times greater than the Earth-Moon distance.

383. (a) West to East.

384. (c) Emerging from total eclipse by the Earth.

385. (a) Approximately equal to d.

386. (b) The Sun is at its farthest distance and the Moon at its nearest.

387. 0.22 AU.

388. (d) June.

389. Mid-March and mid-September 1969.

390. Early June and mid-December 1974.

391. Yes, in the eclipse season late July—early August 1971.

392. (d) June.

393. Late May and mid-November 1975.

394. (a) 173 days.

395. (d) Late October.

396. Late June and mid-December.

397. (d) March.

398. Accurate knowledge of the Moon's orbit.

399. Mid-April.

400. (c) 6 months.

401. Because of the regression (westward motion) of the nodes of the Moon's orbit on the ecliptic at the rate of 360 degrees in 18.6 years or 19.4 degrees per year.

402. Let $\varphi(t)$ be the counterclockwise angle from the meridian plane through the mean Sun to the meridian plane through the observer and let the rotational period of the Earth $P(t) = P_0 + \dot{P}t$, where P_0 is the value of P at $t = 0$ and $\dot{P}$ denotes dP/dt and is assumed constant. Then

$$\frac{d\varphi}{dt} = \frac{2\pi}{P} = \frac{2\pi}{P_0 + \dot{P}t} \approx \frac{2\pi}{P_0}\left(1 - \frac{\dot{P}}{P_0}t\right) .$$

Integrating:

$$\varphi(t) = \varphi_0 + \frac{2\pi}{P_0}\left(t - \frac{\dot{P}}{2P_0}t^2\right) .$$

Let the accumulated discrepancy

$$\delta\varphi = \varphi(t) - \varphi_0 - \frac{2\pi t}{P_0}$$

or

$$\delta\varphi = -\frac{\pi\dot{P}}{P_0{}^2}t^2 .$$

By the statement of the problem, $\delta\varphi = -\pi/4$. Therefore

$$\dot{P} = \frac{P_0{}^2}{4t^2} .$$

Numerically: $P_0 = 86,400$ sec and $t = (2,000)(86,400)(365.26)$ sec so that

$$\dot{P} = 4.685 \times 10^{-13} \text{ sec per sec}$$

or

$$\dot{P} = 1.48 \times 10^{-5} \text{ sec per year}$$

or

$$\dot{P} = 1.48 \text{ milliseconds per century} .$$

403. When the Sun is at its nearest point to the Earth and the Moon at its farthest, the angular diameter of the Moon is less than that of the Sun. Hence the Moon cannot fully eclipse the Sun.

404. One must first calculate the diameter d of the umbral cone at the surface of the Earth (from data for the diameters of the Sun and Moon and their distances from the Earth on the date in question). The maximum duration of a total eclipse is the ratio of d to the speed of the center line of the umbral cone relative to the observer. This relative speed is the eastward orbital speed of the Moon (about 1 km s^{-1}) minus the eastward rotational speed of the observing point on the Earth (0.46 km s^{-1} at the equator and less at any other latitude). For example, if d = 100 km, the maximum duration of totality at an observing point on the equator is about 185 seconds.

405. About 0.54 km s^{-1}. See answer to problem 404.

406. The spot moves from West to East at a speed of about 0.54 km s^{-1}. See answer to problem 404.

407. 370 seconds.

408. (d) 640 years.

409. Primarily because of sunlight scattered by the surrounding uneclipsed atmosphere into the sky above the observer. Lesser contributions come from Earth-light reflected from the Moon and from starlight and airglow.

410. Some matters to be investigated:
     - Credibility and quality of any claimed eyewitness or photographic observations.
     - Detection of any radio signals.
     - Compatibility of claimed motion with Newton's laws and with flight within the atmosphere.
     - Recovered and exhibited parts, materials, or devices unknown on the Earth and documentation thereof.
     - Recovery and exhibition of occupants, dead or alive, and documentation thereof.

The Sky as Observed from the Rotating, Revolving Earth

411. To find
   - phases of the Moon;
   - times of sunrise and sunset;
   - times of moonrise and moonset;
   - dates of vernal equinox, summer solstice, autumnal equinox, and winter solstice;
   - positions of the planets;
   - descriptive tables of solar and lunar eclipses;
   - lists of stellar positions;
   - orbital elements of planetary orbits;
   - motion of planetary satellites;
   and many other astronomical data.

412. (d) The ecliptic.

413. (b) 21 March.

414. All of the stars within an arbitrarily adopted portion of the celestial sphere.

415. Signs of the Zodiac.

416. One for each month of the calendar year.

417. Any one of the twelve constellations of the Zodiac.

418. 12.

419. Polaris, α UMi, is easily visible to the unaided eye and is nearly true North and at an altitude nearly equal to the observer's latitude. At the present epoch Polaris is 0.77 degree from the celestial pole and convenient tables may be used to find true North and latitude exactly from observations of Polaris.

420. (d) Pegasus.

421. October.

422. Southward passage of the Sun through the autumnal equinox (right ascension 12^h, declination zero degrees).

423. (a) 21 March.

424. Right ascension 18^h, declination 23.5 degrees South.

425. All stars in a given constellation are near each other ($\pm$ about 15 degrees) on the celestial sphere but their distances along the line of sight are in general totally unrelated to each other. There are some special exceptions.

426. There is in general no relationship between distances to stars and their angular separations on the celestial sphere.

427. ■ Latitudinal variation of the acceleration due to gravity.
    ■ Wind circulation patterns around high and low pressure areas and in hurricanes and tornadoes (Coriolis force).
    ■ Oblateness of the Earth.
    ■ Motion of a Foucault pendulum.

- Necessary corrections for launching a satellite of the Earth (see problems 120 and 121).
- Corrections required in the targeting of ballistic missiles.
- Diurnal variation of the Doppler shift in the radio signal from a distant spacecraft or in the optical spectrum of a star.

428. Counterclockwise.

429. The mean Sun.

430. (b) Atomic clocks.

431. The sidereal day is the lapse of time between successive meridian transits of the vernal equinox or of any designated star.

432. (a) The vernal equinox is the position on the celestial sphere of the ascending node of the ecliptic on the Earth's equator.
(b) Sidereal time at a specified observing point is the local hour angle of the vernal equinox.

433. (d) Orbital motion of the Earth about the Sun.

434. (Equation of time) = (apparent solar time) − (mean solar time).

435. One radian is the angle subtended at the center of a circle of radius r by an arc of length r along the circumference of the circle. 1 radian = $360/2\pi$ = 57.30 degrees.

436. 5280 feet (1 mile).

437. (a) Ratio of the Earth-Moon distance to the diameter of the Earth.

438. 00^h 07.6^m.

439. One-half of the angular change in the star's apparent position during the course of a year relative to much more distant stars on the celestial sphere; or, stated otherwise, the ratio of 1 AU to the star's distance, usually measured in seconds of arc.

440. A planet moves with respect to the star field from night to night. Also, at a more subtle level, a planet "twinkles" less than a star and has a distinctive color.

441. A personal observation.

442. About 2,500.

443. Select the brightest stars and compare the pattern of their geometrical relationships with a star chart.

444. Because of the brightness of the sky caused by scattered sunlight in the atmosphere.

445. (a) Moonlight scattered by the Earth's atmosphere.

446. About one mile.

447. The circular intersection of the sphere with a plane passing through its center.

448. The circular intersection of the sphere with a plane that does not pass through its center.

449. ▪ Celestial coordinates—Earth-centered, referred to the vernal equinox and Earth's equator (right ascension and declination).
   ▪ Geocentric ecliptic coordinates—Earth-centered, referred to the vernal equinox and ecliptic (longitude and latitude).
   ▪ Heliocentric ecliptic coordinates—Sun-centered, referred to the vernal equinox and ecliptic (longitude and latitude).
   ▪ Geographic coordinates—Earth-centered, referred to Greenwich meridian and equator (longitude and latitude).
   ▪ Topocentric coordinates—centered at a specified geographic point (altitude and azimuth).

   (Analogous systems are used for other planets.)

450. Declination.

451. Right ascension.

452.

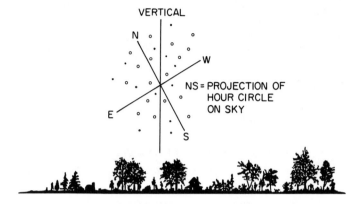

VIEW OF THE SOUTHEASTERN SKY FROM A NORTHERN LATITUDE SITE

453. 30 degrees.

454. The culmination of a star occurs at the moment at which it has its greatest altitude, i.e., as it crosses the observer's meridian. Circumpolar stars have both upper (greatest altitude) and lower (least altitude) culminations.

455. The latitude is determined by the described observations without the need for any other data. However, determination of longitude requires knowledge of the Greenwich Mean Time of the observation. The latter point is obvious from the fact that the same set of three altitudes occurs at observing positions progressively westward at the same latitude as that of the original one, as the Earth rotates.

456. (d) On a small circle centered on the substellar point.

457. Tropic of Cancer.

458. 00^h (midnight).

459. 3 degrees.

460. 58 degrees.

461. 24.8 degrees.

462. (b) Latitude.

463. 84 minutes.

464. At any point on the equator.

465. (d) Local sidereal time.

466. (a) 35.5 degrees (maximum) (azimuth 180 degrees);
     (b) 11.5 degrees (minimum) (azimuth zero degrees).

467. For a northern hemisphere observer, the altitude of any star in the southern sky has its maximum value as the star transits the observer's meridian. At this moment, the star is said to be at culmination and it is true South (azimuth 180 degrees). True North (azimuth zero degrees) is in the opposite direction.

468. 23 degrees South.

469. (a) Fly westward at 1,700 km hr^{-1}.

470. (b) The equator.

471. The interval of time between successive vernal equinoxes.

472. (d) Gregorian calendar.

473. (d) 1996.

474. (c) 1976. The century year 2100 is not a leap year.

475. Eight years.

476. (b) Increases by about fifty minutes.

477. (e) Star, Sun, Moon.

478. Because of the tilt (23.5 degrees) of the Earth's rotational axis to the normal to the ecliptic plane.

179. (c) Tilt of the rotational axis of the Earth to the normal to the ecliptic plane.

180. (d) September.

181. (c) The Sun's rays strike the Earth at smaller angles to its surface in the winter.

182. (a) There would be no seasons.
(c) The equation of time would be zero.
(d) The average temperature of the Earth would be about the same as it now is.

183. (c) The rotational axis of the Earth were perpendicular to the ecliptic.

184. (c) Draco.

185. The two points are diametrically opposite, at latitudes 66.5 degrees North and 66.5 degrees South (i.e., on the Arctic and Antarctic Circles), respectively.

186. 66.5 degrees North (on the Arctic Circle).

187. 28.5 degrees.

188. (b) Declinations greater than 49 degrees North.

189. (b) 90 degrees − φ.

190. 22 September. The beginning of the winter night is the date of each year on which the Sun ceases to rise above the horizon at any time during the 24-hour day. (Atmospheric refraction neglected.)

191. $24^h 50^m$.

192. (a) $24^h 50^m$.

493. 14.3 degrees.

494. 41.5 degrees.

495. Azimuth = 270 degrees; altitude = zero degrees.

496. 42 degrees North or more exactly 42 ($\pm$0.8) degrees North.

497. 1.0 degree.

498. (c) Local apparent solar time.

499. ■ Sundials are the most ancient and obvious astronomical instruments for indicating the passage of time. The fundamental simplifying principle of a sundial is as follows. If the style, the shadow-casting edge of the gnomon, points at the celestial pole (in either hemisphere), it is parallel to the Earth's axis of rotation and lies in the vertical plane through the site that is coincident with the local meridian plane. Hence the Sun and all other celestial objects having a given local hour angle t lie in a plane hinged along the style. This plane is the same irrespective of the declinations of the objects and makes a dihedral angle t to the local meridian plane.

■ In the most common form of a sundial, the shadow of the style falls on a horizontal plane, with the center of the dial at the foot of the style. A specific hour-angle plane intersects the horizontal plane along the line through the foot of the style that makes an angle β to the horizontal North-South line such that

$$\tan \beta = \tan t \sin \varphi \,.$$

At a given latitude φ, β is independent of declination as noted above and is therefore the same for any specific local apparent solar time on any date of the year. No other orientation of the style yields this result.

Two special cases are readily visualized.

■ At $\varphi = 90$ degrees, the style is vertical and hour angles ($\pm 12^{\text{h}}$) are equally spaced around a full circle in the horizontal plane.

At $\varphi = 0$ degree (the equator) the style may be visualized as a horizontal North-South wire whose shadow falls along parallel North-South lines on a horizontal surface, the lines being unequally spaced. Or if the shadow falls on a semi-cylindrical surface whose axis is the style, the North-South indicator lines are equally spaced. Such a semi-cylindrical arrangement is applicable to any other latitude.

An example of the more common horizontal sundial at an intermediate latitude $\varphi = 40$ degrees North, $\sin \varphi = 0.6428$, is as follows:

Local Apparent Hour Angle, t	β
$\pm 9^h$, 135 degrees	± 147.3 degrees
8, 120	131.9
7, 105	112.6
6, 90	90.0
5, 75	67.4
4, 60	48.1
3, 45	32.7
2, 30	20.4
1, 15	9.8
0, 0	0.0

500. The Earth is a huge gyroscope whose axis tends to point at a fixed point on the celestial sphere. But because of its oblate shape and because the planes of the orbits of both the Sun and the Moon are inclined to the Earth's equator, there is an average tidal torque at right angles to the Earth's rotational axis (principally by the Moon). The result is a counterclockwise precession of the Earth's axis on the star field in a circle of radius 23.5 degrees centered on the pole of the ecliptic.

501. Precession of the equinoxes.

502. (c) Gyroscopic precession of the axis of rotation of the Earth.

503. (d) 25,800 years.

504. 43 degrees.

505. 2,150 years.

506. By about the year 2600, the vernal equinox will have precessed westward and, by a lesser amount, southward on the star field from the constellation Pisces into the constellation Aquarius.

507. Sirius.

508. (b) Cassiopeia.

509. (a) Orion.

510. (a) Taurus; (b) Orion; (c) Canis Major; (d) Auriga.

511. Taurus.

512. Markab.

513. (a) Pegasus.

514. (d) Cygnus.

515. (d) Deneb.

516. (a) Ursa Minor; (b) Lyra; (c) Auriga; (d) Taurus.

517. (d) Aquila.

518. A personal observation.

519. (b) The brightest star in Orion.

520. (d) A constellation that contains the star Deneb.

521. (a) Pegasus; (b) Lyra; (c) Aquila; (d) Boötes.

522. Betelgeuse and Rigel.

523. Castor and Pollux.

524. A personal observation, possibly projected backward or forward in time. Or, note that the right ascensions of the two pointer stars are about 11^h.

525. (d) Sirius.

526. Taurus.

527. Virgo.

528. A personal observation.

529. A personal observation.

530. Vega is the brightest star in the constellation Lyra. On 15 September it crosses an observer's meridian at about 19^h standard time (20^h daylight time) at an altitude from the south point of 129 degrees minus the observer's latitude. Vega is the northwesterly apex of the "summer triangle", Deneb-Vega-Altair.

531.

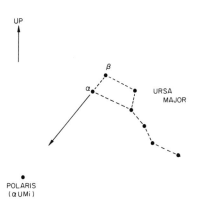

VIEW OF THE NORTHERN POLAR SKY —
EARLY EVENING, LATE MARCH

As illustrated, a line through the two second magnitude stars β and α UMa, the pointer stars of the Big Dipper, passes near Polaris (α UMi), also a second magnitude star, and the northern celestial pole. The angle between α UMa and the pole is 5.3 times the angle (5.4 degrees) between β and α UMa. Polaris is about 0.77 degree from the pole. The meridian plane through the observer defines the vertical North-South plane.

532. Personal observations.

533. Aldebaran.

534. Gemini.

535. (d) Aquila.

536. (b) Cygnus.

537. (c) Taurus.

538. (c) A cluster of stars in the constellation Taurus.

539. Orion.

540. (a) Auriga; (b) Boötes; (c) Orion; (d) Aquila.

541. (d) Cassiopeia.

542. Two of the following: Ursa Major, Ursa Minor, Auriga, Perseus, Cassiopeia, Cephus, Draco.

543. (c) Orion.

544. Vega, Deneb, and Altair.

545. A personal observation.

546. 00^h (local midnight).

547. 82.6 degrees latitude South.

548. (c) Lie on a small circle on the celestial sphere centered on the observer's zenith.

549. From latitude 37.3 degrees North to 90 degrees South.

550. At latitude 38 degrees 47 minutes North.

551. Latitude 50 degrees South.

552. 5 km s^{-1} toward the Sun.

553. The full Moon is (approximately) diametrically opposite to the Sun on the celestial sphere. Hence, when the Sun is at its most southerly declination (21 December), the full Moon is at its most northerly declination.

554. 40 degrees South.

555. Longitude 165 degrees East.

556. (d) Is high in the northeastern sky in the early evening.

557. 14^h 32^m.

558. Latitude 54 degrees North; longitude 30 degrees East.

559. 20^h on Sunday.

560. 03^h on Friday.

561. (a) 18^h (4 minutes earlier each day).

562. Longitude 90 degrees West.
     Latitude 33 degrees North.

563. (a) 37 degrees.

564. 43 degrees.

565. Latitude 50 degrees North. (The declination of the Sun is zero and its meridian transit altitude is 40 degrees.)

566. 07^h 12^m.

567. (c) 06^h.

568. Late October to mid-February.

569. About the 20th of August.

570. (a) 90 degrees West.

571. 06^h.

572. (a) 285 degrees.

573. (a) Cover the lens for, say, 5 minutes at the end of the first half-hour of the exposure, thus dividing the trail into two unequal segments and establishing the sense of rotation of the star field.
(b) On an enlarged print find the center of curvature of the arcs of the star trails using a simple transparent overlay having circles of various radii. Then measure the average angle θ at the center of the arcs of several star trails. The sidereal period of revolution of the Earth is $(360/\theta)$ (exposure time).

574. 06^h 43^m.

575. Mid-April, if in the proper range of latitude, of course.

576. 16.5 degrees.

577. 01^h (15 degrees).

578. (a) 310 degrees.

579. (c) December.

580. (d) The Sun is in Aries. (False)

581. 12^h GMT.

582. On the date of the winter solstice, the center of the Sun's disc is on the horizon there at local noon, neglecting atmospheric refraction.

583. Standard Time itself is an artificially constructed system. Averaged over a time zone, it yields approximately equal values of the time intervals between sunrise and noon and between noon and sunset, whereas Daylight Saving Time yields a one-hour lesser value for the former interval and a one-hour greater value for the latter interval.

   Local apparent solar time (sundial time) is the most "natural" system to use for activities related to sunlight but such time varies continuously with longitude at any moment and is therefore so inconvenient as to be effectively useless for an advanced civilization.

584. Autumnal equinox. Note that the local sidereal time is equal to the hour angle of the vernal equinox, whereas the local solar time is equal to the hour angle of the Sun plus 12^h.

585. (c) Autumnal equinox. (See problem 584.)

586. Mid-December.

587. Latitude 60 degrees South.

588. (a) The altitude h of Polaris.

589. Declination 20 degrees South.

590. 09^h 34^m.

591. A personal observation.

592. (d) 24^h (or 00^h).

593. (a) 01^h.

594. (d) 12^h.

595. (c) Approximately 19^h 26^m.

596. (d) 240 degrees.

597. 18^h.

598. About 00^h 26^m.

599. (a) 13^h 24^m.

600. 18^h.

601. About 11 days.

602. (a) The meridian of the Fiji Islands.

603. (b) Longitude 75 degrees West.

604. 21^h 16^m CDT.

605. (c) 90 degrees $- \varphi$.

606. 122 degrees.

607. (a) 08^h Thursday.

608. (b) 90 degrees.

609. 22^h Wednesday.

610. Longitude 30 degrees East.

611. 10^h 36^m.

612. ■ If local apparent solar time (sundial time) is used, the latest sunrise, the earliest sunset, and the shortest interval between sunrise and sunset all occur at the winter solstice; and likewise the earliest sunrise, the latest sunset, and the longest interval between sunrise and sunset all occur at the summer solstice.

■ Thus, it is evident that the facts stated in the problem are attributable to the adoption of mean solar time for civil time keeping. The mean sun is a fictitious point that moves along the celestial equator at a uniform rate calculated to be equal to the average rate at which the apparent (actual) sun completes its annual motion along the ecliptic. Because of the eccentricity of the Earth's orbit and the 23.5-degree inclination of the ecliptic to the celestial equator, the mean sun crosses an observer's meridian earlier than does the apparent sun during parts of the year and later during the other parts. The effect is summarized by a quantity called the equation of time E, where

$$E = \text{Apparent Solar Time} - \text{Mean Solar Time} ,$$

usually measured in minutes and seconds of time. The values of E as a function of calendar date are essentially the same for any year, lying in the range $+16^m$ to -15^m.

■ For example, on 9 December $E = +8^m$ and on 4 January $E = -5^m$. This 13-minute decrease in E combined with the annual variation of the declination of the apparent sun accounts for the facts stated in the problem.

■ The following diagrams show the approximate times of sunrise and sunset at latitude 44 degrees North, near the winter solstice, in both local apparent solar time and local mean solar time. The curves for the former are symmetric about the winter solstice but the rapid variation of the equation of time during this season causes the curves for the latter to be markedly asymmetric.

■ The curves are essentially the same for any year.

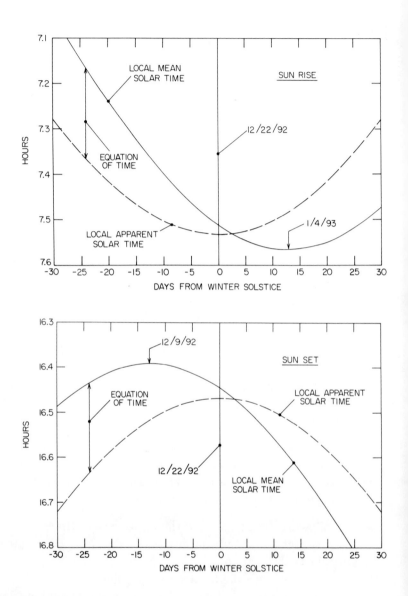

613. (a) Longitude 15 degrees East.

614.   GHA (object) = GHA γ − α (object) .

If the result in degrees is negative, add 360 degrees; if greater than 360 degrees, subtract 360 degrees. Also note that the side-real hour angle (SHA) of a star is equal to (360 degrees − α).

615. 01^h Tuesday.

616. 20^h 33^m.

617. (c) 03^h on 23 January.

618. (a) Longitude 37.5 degrees West.

619. The ship's latitude can be determined by measuring the altitude of the Pole Star (Polaris) with a marine sextant or, to lesser accuracy, with a primitive device such as a protractor and plumb line. Alternatively, the latitude can be determined by measuring the altitude of any star (or planet or the Sun or the Moon) at its culmination if the declination of the object is known. Also, either an observation of Polaris or an object at its culmination in the southern sky determines the ship's heading relative to true North. Hence, the ship can be guided from a point in Europe to a point at about the same latitude in North America without the use of any timepiece or any intermediate knowledge of longitude. There are obvious modifications of the technique for sailing to a latitude different than that of the point of departure.

620. ■ Observed altitudes (ideally simultaneously) at a known Greenwich Mean Time and date of at least two celestial objects (Sun, Moon, planets, or stars) having known right ascensions and declinations. It is noted that the use of only two celestial objects yields a twofold ambiguity in position. However, the two geometrically possible positions are usually much different and can be distinguished by auxiliary knowledge. Observation of three or more objects removes the ambiguity.

■ Observation of the altitudes of the same object at two different times also yields position, to within twofold ambiguity, if the observer is at a fixed point or corrects for estimated change in position. Observation of both the altitude and azimuth of a single object (e.g., the Sun) at a known GMT also provides the basic information for determination of position.

■ In one modern form of "celestial" navigation, sophisticated artificial satellites supplant natural astronomical objects and the need for a chronometer.

621. (a) It is essential to also know Greenwich Mean time (GMT).

622.

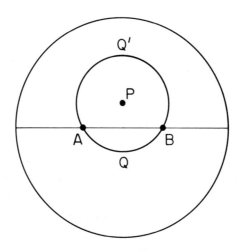

Without loss of generality, the great circle on which the two points A and B lie may be visualized as the equator of a sphere, as indicated in the above diagram. Then imagine a small circle AQBQ' on the surface of the sphere centered around P and passing through both A and B. The great circle path from A to B is an excursion in longitude only, at zero latitude. In contrast, the small circle path from A to B via Q spans the same range of longitude but also makes an excursion in latitude. Hence the small circle path AQB is longer than the great circle path AB. This is a general result for any small circle passing through both A and B.   Q.E.D.

Other Planets, Their Satellites, and Rings

623. Neptune.

624. (d) Inversely as the fourth power of its distance.

625. $(10^{-2})(270) = 2.7$ degrees Kelvin.

626. The classification is on the basis of mean density. The terrestrial planets (high mean densities and solid surfaces) are Mercury, Venus, Earth, and Mars. The Jovian planets (low mean density and mainly gaseous and liquid) are Jupiter, Saturn, Uranus, and Neptune. Pluto has a low mean density but has a solid surface. Minor planets (asteroids) are terrestrial in nature.

627. (d) Mars.

628. Mercury, Venus, Mars, Jupiter, and Saturn.

629. Jupiter, Saturn, Uranus, and Neptune.

630. Venus.

631. (a) Gravitational attraction of every element of the planet by every other element.

632. ■ For an isolated body of fluid, a sphere is the shape of minimum gravitational potential energy. (If the fluid is rotating, an oblate spheroid is the shape of minimum potential energy.)

■ An isolated solid body also tends toward a spherical shape because of self-gravitational attraction, but will not assume that shape if the crushing strength of the material exceeds the gravitational pressure. Consider a cylindrical body of radius a, length 4a, mass m, density ρ, and crushing strength C. For an approximate analysis, imagine the body to be composed of two sub-cylinders, each of length 2a. Then the gravitational attraction between the two is approximately

$$\frac{Gm^2}{16a^2} = G\pi^2 a^4 \rho^2 . \tag{1}$$

The crushing pressure across the dividing plane between the two sub-cylinders is

$$\pi a^2 \rho^2 G . \tag{2}$$

The threshold condition for crushing to occur is

$$\pi a^2 \rho^2 G = C \tag{3}$$

or

$$a = \frac{\sqrt{C/\pi G}}{\rho} . \tag{4}$$

■ For rocky material, a representative value of C is 1×10^9 dyne cm^{-2} and of ρ is 2.8 g cm^{-3}. Using (4), one finds that if a exceeds 250 km, the body will tend toward sphericity, whereas if a is less than 250 km, it can retain a markedly nonspherical shape. For cometary material, the threshold value of a is perhaps an order of magnitude less than that for rocky material. This treatment yields only a crude estimate of a but does illustrate the nature of the problem.

633. ▪ Suppose that the mountain is a solid rigid cone having a circular base of radius a, height a, mass m, and mean density ρ. The mass of the mountain is

$$m = \frac{\pi a^3 \rho}{3} \qquad (1)$$

and its weight is

$$\frac{\pi a^3 \rho g}{3} . \qquad (2)$$

The weight is spread over an area πa^2. Hence, the threshold crushing condition is

$$\frac{\pi a^3 \rho g}{3 \pi a^2} = C , \qquad (3)$$

where C is the crushing strength of the underlying crust of the planet. From (3)

$$a = \frac{3C}{\rho g} . \qquad (4)$$

▪ Representative values of C and ρ for natural rocky materials are 1×10^9 dyne cm^{-2} and 3 g cm^{-3}, respectively. With these values and g = 982 cm s^{-2}, it is found by (2) that for the Earth

$$a = 10 \text{ km} . \qquad (5)$$

This simple analysis yields, of course, only a crude estimate of the maximum possible height of a mountain on a planet, but it does illustrate the nature of the problem.

▪ By (4) it is noted that the upper limit value of a is inversely proportional to g. For the Moon and the terrestrial planets Mercury, Venus, the Earth, and Mars, the surface values of g are 162, 363, 860, 982, and 374 cm s^{-2} and by (4), the corresponding threshold values of a are 61, 21, 11, 10, and 26 km, respectively. The heights of the highest actual mountains on these bodies are as follows: 10, 7, 11 (Maxwell Montes), 9 (Mt. Everest), and 27 (Olympus Mons) km, respectively.

634.

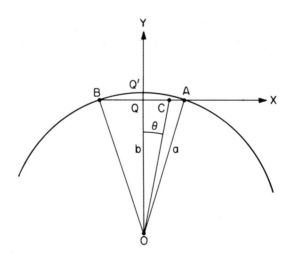

■ In the diagram, A represents New York; B, San Francisco; the arc AQ'B, a great circle between A and B; AQB, the tunnel along the X-axis with midpoint at Q; O, the center of the Earth, radius a (6372 km); and C, the position of the car at a particular moment.

■ The respective latitudes and longitudes of New York and San Francisco are 40.75 degrees North, 73.99 degrees West; and 37.78 degrees North, 122.41 degrees West. The great circle, sea level distance between the two cities is 4,130 km and the straight line distance is 4,057 km. The angle AOB = 37.13 degrees, b = OQ = 6,040 km, and QQ' = 332 km, the depth of the tunnel at its midpoint Q.

■ The inward radial acceleration due to gravity g is a few percent less at Q than at the surface (see problem 65). In the following analysis, however, the variation of g with x is neglected and its sea level value 9.82 m s^{-2} is adopted.

At C, the x-component of the gravitational force on the car of mass m is

$$F_x = -mg \sin \theta \ , \tag{1}$$

or

$$F_x = -\frac{mg}{b} x \ , \tag{2}$$

approximately. Equation (2) is recognized as being the equation of a single harmonic oscillator with period

$$P = 2\pi\sqrt{b/g} , \qquad (3)$$

as for a simple pendulum of length b. P is about 82 minutes. The one-way trip time, New York to San Francisco, is P/2 = 41 minutes.

■ The car passes the midpoint Q at a speed of about 2.55 km s^{-1} (5,700 miles per hour). It comes to rest at the San Francisco mouth of the tunnel and, if not locked in place there, coasts back to New York, then back to San Francisco, etc.

635. Because of the gravitational attraction of every element of the body on every other element, in other words the weight of the overhead material.

636. The mean density, which is the ratio of the total mass to the total volume.

637. (c) Its internal composition.

638. (b) Mean densities.

639. (d) Saturn.

640. Jupiter, Saturn, Uranus, Neptune, and Pluto.

641. Saturn.

642. (c) The reported data are incompatible with Kepler's and Newton's laws. (False)

643. (c) Between the orbits of Saturn and Uranus.

644. Tidal torques on nonspherically symmetric bodies and internal dissipation of energy.

645. (c) H_2 (14 degrees Kelvin).

646. The temperature of its atmosphere, the gravitational escape speed from its surface, and the molecular weight of the atmospheric gas.

647. (c) The gravitational escape speed from its surface.

648. Oxygen.

649. 64 [2^6].

650. Hydrogen, nitrogen, oxygen, and carbon.

651. Gamma.

652. H_2O, NH_3, CH_4 (marginal), and CO_2.
(Estimated temperature of a satellite of Saturn $\approx 270/\sqrt{10} =$ 85 degrees Kelvin)

653. The sunlit surface temperatures of Io, Ganymede, and Titan are much less than the sunlit temperature of the Moon. Hence, molecular speeds are also much less and gases escape much less rapidly than from the Moon.

654. (a) Jupiter.

655. Jupiter, Saturn, Uranus, and Neptune.

656. Earth, Jupiter, Saturn, Uranus, and Neptune.

657. (b) The breakup of objects in close orbits about a planet or the Sun.

658. (b) Tidal forces.

659. (a) Tidal forces.

660. ■ The principle of the Roche limit is described and an approximation to its magnitude is found as follows:

A hypothetical object consists of two rigid, solid spheres, each of radius r, mass m, and mean density ρ_1. Imagine that these two spheres are in contact and are in a circular orbit of radius R about a spherical planet of radius a, mass M, and mean density ρ_2 with R > a $\gg$ r. Further imagine that the line through the centers of the two spheres passes through the center of the planet.

The mutual gravitational attraction between the two spheres tends to hold them together whereas the "tidal force" tends to pull them apart. The Roche limit is derived from the condition that these two opposing forces be of equal magnitude.

For the center of mass of the two spheres (their point of contact), the Newtonian orbital equation (Kepler's third law) gives

$$\omega^2 = \frac{GM}{R^3} \tag{1}$$

where ω is the angular speed in orbit of the composite object.

The gravitational attraction between the two spheres is:

$$\frac{Gm^2}{4r^2} . \tag{2}$$

In the reference system rotating with angular speed ω the downward dynamical force on the lower sphere is

$$\frac{GMm}{(R-r)^2} - m(R-r)\omega^2$$

and the upward dynamical force on the upper sphere is

$$-\frac{GMm}{(R+r)^2} + m(R+r)\omega^2 .$$

With the help of (1), each of these forces is found to be equal to

$$3m\,r\,\omega^2, \tag{3}$$

the so-called tidal force.

The Roche limit corresponds to the equality of (2) and (3), namely

$$\frac{Gm^2}{4r^2} = 3m\,r\,\omega^2. \tag{4}$$

Using (1) and the facts that $M = 4\pi\,a^3\rho_2/3$ and $m = 4\pi\,r^3\rho_1/3$, equation (4) becomes

$$\frac{R}{a} = 2.3\left(\frac{\rho_2}{\rho_1}\right)^{1/3}. \tag{5}$$

■ The more rigorous treatment of Roche for a fluid satellite gives the same functional form as (5) but yields a slightly different numerical factor, namely:

$$\frac{R}{a} = 2.5\left(\frac{\rho_2}{\rho_1}\right)^{1/3}. \tag{6}$$

Inasmuch as the one-third power of the ratio of the two mean densities is usually near unity for natural systems, (6) is often written simply as

$$R = 2.5\,a\,. \tag{7}$$

The implication of (7) is that a satellite having zero tensile strength cannot exist in an orbit having $R < 2.5\,a$ but will be disrupted by "tidal tension".

Artificial satellites maintain their integrity for R less than 2.5 a by virtue of their tensile strength as do small natural bodies, such as those in planetary rings.

661. ■ For planets having one or more satellites, the quantity GM, where G is the universal gravitational constant and M the mass of the planet, occurs in Kepler's third law:

$$P^2 = \frac{4\pi^2}{GM}\,a^3\,.$$

The period of revolution of a satellite is P and the semimajor axis of its elliptical orbit is a; both P and a are observable. Then, GM can be calculated from the above equation; and

$M = GM/G$ where the value of G is provided by the Cavendish technique.

■ The value of GM for a planet that has no natural satellites can be determined by the observed magnitude of the perturbation that it exerts on the orbits of other planets or asteroids.

■ In recent years, the most accurate values of planetary masses have been derived from the trajectories of flyby and orbiting spacecraft.

662. See answer to problem 661.

663. (a) Analysis of the gravitational perturbation that it produces on the orbital motions of Venus and other planets.

664.

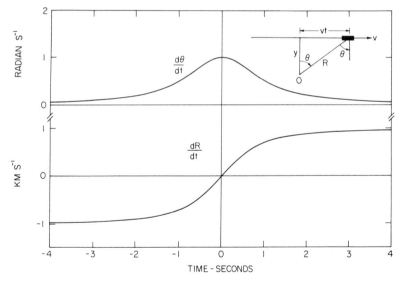

As shown in the insert in the above diagram

$$R^2 = y^2 + (vt)^2 , \text{ and} \tag{1}$$

$$\frac{dR}{dt} = \frac{v^2t}{\sqrt{y^2 + (vt)^2}} ; \tag{2}$$

$$\theta = \text{arc tan } \frac{vt}{y} , \text{ and} \tag{3}$$

$$\frac{d\theta}{dt} = \frac{v/y}{1 + (vt/y)^2} \cdot \qquad (4)$$

Plots of dR/dt vs t and $d\theta/dt$ vs t are shown in the diagram for the parameters specified in the statement of the problem.

665. ■ In April 1965, Pettengill and Dyce directed a powerful, pulsed radar beam at Mercury, using the Arecibo radio telescope. In each of four runs during a 20-day period near the date of inferior conjunction they measured, in effect, the difference (without knowledge of algebraic sign) between the Doppler frequency shifts of pulses reflected from surface patches at known distances on opposite sides of the center of the disc of the planet. It is evident that such measurements provide information on the rotational rate of the planet, e.g., the left side of the planet moving toward the observer and the right side moving away from the observer, or vice versa.

The geometric basis of such work is shown in the first diagram.

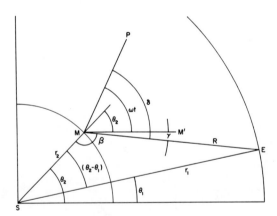

■ The Sun, Mercury, and the Earth are labeled S, M, and E, respectively. At $t = 0$, all three objects lie along the horizontal axis. The line MM' is parallel to that axis, thereby representing a fixed sidereal reference line.

For the purpose of exemplary calculations, approximate the orbits of Mercury and the Earth as coplanar circles. Then

$$\theta_1 = \frac{360}{365.256} t = 0.986\, t$$

$$\theta_2 = \frac{360}{87.969} t = 4.092\, t$$

with θ_1 and θ_2 in degrees and t in days; and

$$r_1 = 1.000 \text{ AU}$$

$$r_2 = 0.387$$

Suppose that line MP, fixed in the planet, lies along the Sun-Mercury-Earth line at the moment of inferior conjunction, t = 0, and suppose that it is rotating at the sidereal angular rate ω degrees per day. Take ω to be positive for counterclockwise rotation (prograde) and negative for clockwise rotation (retrograde).

As shown in the diagram:

$$R^2 = r_1{}^2 + r_2{}^2 - 2r_1r_2\cos(\theta_2 - \theta_1) \qquad (1)$$

$$\sin\beta = (r_1/R)\sin(\theta_2 - \theta_1) \qquad (2)$$

$$(\theta_2 - \theta_1) = 3.106\, t \qquad (3)$$

$$\gamma = 180 - \beta - \theta_2 \qquad (4)$$

$$\delta = \gamma + \omega t . \qquad (5)$$

The apparent rate of rotation of the planet is

$$\frac{d\delta}{dt} = \frac{d\gamma}{dt} + \omega$$

or

$$\frac{d\delta}{dt} = \omega - \frac{d\beta}{dt} - \frac{d\theta_2}{dt} . \qquad (6)$$

Equation (6) is used for numerical examples for several values of ω, viz.:

$$\omega = \frac{360}{59} = 6.102 \text{ degrees day}^{-1} \, ,$$

$$\omega = \frac{360}{87.969} = 4.092 \, , \text{ and}$$

$$\omega = -\frac{360}{44.4} = -8.108 \, .$$

The theoretical expectations are shown in the second diagram.

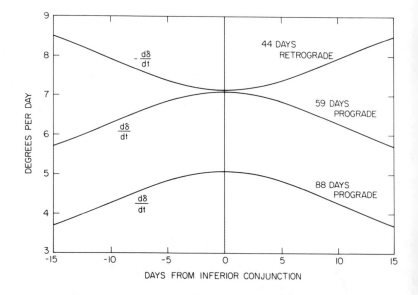

- The observations of Pettengill and Dyce within a few days of inferior conjunction were compatible with either prograde rotation with a 59-day sidereal period or retrograde rotation with a 46-day period (cf. two upper curves in the diagram) *but* were grossly incompatible with the previously accepted (synchronous) value of 88 days (lowest curve). Their fourth run about 15 days after inferior conjunction resolved the former ambiguity and conclusively established 59 (± 5) days as the sidereal period of Mercury's prograde rotation. Subsequent observations and theoretical considerations yield the more accurate value of 58.65 days, almost exactly two-thirds of the period of revolution.

666. See answer to problem 665.

667. 176 days, twice the period of revolution and three times the period of rotation. This is a noteworthy case of a 3 to 2 tidal spin-orbit resonance.

668. 51 degrees Kelvin.

669.

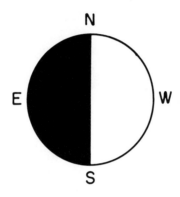

670. As specified, a particular side of Venus faces the Sun every 117 days, and inferior conjunctions occur every 584 days. The latter period is almost exactly 5.0 times the former. Hence Venus makes five retrograde rotations relative to the Sun during each synodic period and therefore presents the same face to the Earth at every inferior conjunction.

671. (c) 117 days.

672. (a) Microwave radiometers.

673. (a) 90 times as great as that at the surface of the Earth.

674. By the limb-to-limb difference of the Doppler shift of reflected pulses from a radar. See answer to problem 665.

675. 103 days.

676. (b) Earth-based radar telescopes.

677. By observation of the rotational movement of distinctive surface features visible with an optical telescope.

678. (b) 23 times greater than at conjunction, assuming circular orbits.

679. (a) CO_2.

680. (c) A thin covering of frozen carbon dioxide and water.

681. (d) Most craters appear to be of meteoric origin.

682. (c) Temperature of its interior. (False)

683. 11.1 hours.

684. (d) The surface of Mars.

685. (c) Established the absence of complex organic molecules in surface soil.

686. Personal assessment in the light of present knowledge.

687. The tidal force by Jupiter on the Earth at opposition is only $1/169,000^{th}$ of that by the Moon and only $1/77,600^{th}$ of that by the Sun. Moreover, the combined lunar and solar tidal forces on the Earth vary with the phase of the Moon and its orbital distance from the Earth by over 50%. Hence the "Jupiter Effect" is enormously implausible.

688. The planet must consist almost entirely of light elements, hydrogen and helium, under the high pressure of self-gravitational forces.

689. (b) It is the largest.

690. (a) Period of revolution of the satellite.

691. (d) None of the above methods.

692. (a) In situ magnetic field measurements near the planet.

693. (a) 1,400.

694. (d) 2.6×10^6 km.

695. 27.0 m s^{-2} (i.e., 2.75 g).

696. 2.63.

697. (d) Its atmosphere contains complex turbulent cloud systems.

698. (b) Jupiter has an internal heat source of the same nature as that mainly responsible for the Sun's heat. (False)

699. ▪ Jupiter has sixteen known satellites. Four of these—Io, Europa, Ganymede, and Callisto—were discovered by Galileo in 1610 and are easily observed with binoculars. Their radii range from 1570 to 2630 km and their mean densities diminish with increasing orbital radii from 3.57 to 1.86 g cm^{-3}. These four Galilean satellites plus four recently discovered, much smaller ones at lesser radial distances, are in nearly circular prograde orbits in the planet's equatorial plane.

▪ Beyond the orbit of Callisto (semimajor axis 26.4 Jovian radii), there are eight small satellites in orbits that are markedly eccentric and inclined at large angles to the planet's equatorial plane. Their semimajor axes range from 155 to 330 Jovian radii.

▪ The four outermost satellites are in retrograde orbits whereas the twelve others are in prograde orbits. It is thought that the inner, regular group of eight satellites was part of the process that created the planet itself but that the outer eight may be asteroids, captured later in astronomical history.

700. Io, the innermost Galilean satellite of Jupiter.

701. (c) Has no effect on its orbital motion.

702. (c) Ices of ammonia, carbon dioxide, and water with an admixture of heavier elements.

703. Jupiter.

704. (b) They revolve in retrograde orbits.

705. (a) Titan.

706. 5.9 years.

707. (c) Pioneer 11.

708. (b) Titan.

709. (b) Pioneer 11.

710. (a) Observation of the radial dependence of the Doppler shift of absorption lines in reflected sunlight.

711. (a) Consist of large and small pieces of ice.

712. It is evident that a final state in which all elements of the ring system revolve in circular orbits in the equatorial plane of the planet is the most durable that one can imagine inasmuch as (ideally) no collisions then occur.

   The detailed way in which this state is approached involves inelastic collisions in which the resulting fragments (or composite bodies) have a resultant linear momentum intermediate in direction between the momentum vectors of the two colliding objects. Hence, there is a tendency toward a common orbital plane and toward circular orbits. The equatorial plane of an oblate planet is the uniquely favored one because orbits of different radii and eccentricities in any other plane precess with respect to each other and further collisions occur.

713. (c) Titan has a dense atmosphere, principally of nitrogen.

714. (a) Roche (see answer to problem 660).

715. (c) The orientation of its rotational axis.

716. The Sun crosses the equatorial plane of Uranus in a nearly North-South or South-North direction at 42-year intervals. Halfway between these two equinoxes the Sun is almost directly over the southern pole of the planet and 42 years later almost directly over the northern pole.

717. (c) 42 years.

718. (c) 42 years.

719. Observing, with an optical photometer attached to a telescope on an aircraft, the occultation of stars as Uranus moved across the star field.

720. 120 AU.

721. (c) Tombaugh, on 18 February 1930.

722. An early value of Pluto's radius was about 3,000 km, based on an estimate of the angular diameter of its very small image formed by an optical telescope. Some years later, Pluto's failure to occult a particular star indicated that 3,000 km was an upper limit on its radius.

The brightness of Pluto is a second basis for an estimate of its radius. If its surface albedo is assumed to be 1.0, its radius is about 1,000 km, a lower limit; if 0.5, 1,400 km, etc.

The most precise present value is $1,142 \pm 21$ km, a product of the study of the planet's occultation of its recently discovered satellite, Charon.

723. (c) 1/1,600.

724. (b) Measuring its brightness.

725. (d) Measuring the cyclic variation of its brightness.

726. Charon.

727. By observing the cyclic variation of its brightness.

728. See answer to problem 727.

729. In descending order of size with radii in parenthesis: Ganymede (2,631 km), Titan (2,575 km), Mercury (2,439 km), Triton (1,900 km), Moon (1,738 km), Pluto (1,125 km).

Asteroids, Comets, and Meteoroids

730. (b) A minor planet.

731. A minor planet. Most of the known asteroids are in more or less circular orbits near the plane of the ecliptic with semimajor axes (median value 2.7 AU) between those of Mars and Jupiter.

732. (c) Size.

733. Over 3,000.

734. See answer to problem 731.

735. (a) 4.4 years.

736. (b) 2.9 years.

737. Principally by its movement relative to the star field. Also by characteristic spectral modification of reflected sunlight and by cyclic variation in brightness, the latter because of rotation of a nonspherical body having perhaps also a variable albedo over its surface.

738. (c) Juno.

739. (a) An asteroid.

740. (b) Europa. (False)

741. 5.67 m s^{-1}.

742. The gravitational field of an asteroid is too weak to retain an atmosphere. Stated otherwise, the gravitational escape speed is less than the typical speed of a molecule of gas.

743. 1.25 pounds.

744. 0.20. The fraction of sunlight reflected is equal to the albedo A and the fraction absorbed and then emitted in the infrared is $(1 - A)$. In this problem $(1 - A)/A = 4$.

745. (d) A special class of asteroids.

746. (c) Illustrate a special solution of the three-body gravitational problem.

747. (b) Are near the Lagrangian points L_4 and L_5 relative to the Sun and Jupiter.

748. By observing the cyclic variation of its brightness.

749. (b) Observing cyclic variations in their brightness.

750. There is substantial geological evidence that a large asteroid struck the Earth at that time, presumably causing a cloud of dust and smoke in the atmosphere that persisted for many years and caused a marked cooling of the Earth's lower atmosphere and surface. A large, buried crater in the Yucatan Peninsula of Mexico is tentatively identified as the impact site.

751. An asteroid is an inert rocky or metallic body whereas a comet consists of frozen light compounds (H_2O, CO_2, CH_4, etc.) and inert dust such that a mixture of gas and dust is emitted from its nucleus as it approaches the Sun and is heated by sunlight.

752. 2 million years.

753. (b) 2.0×10^6 years on its inbound flight to perihelion.

754. (c) The orbit of a comet can be determined by observing its motion on the star field and using Kepler's laws.

755. Kepler's laws are assumed to be applicable to its motion.

756. (c) Have prograde orbits. (Comet Halley in a retrograde orbit is a noteworthy exception.)

757. (d) The orbit of a comet can be determined by observing its motion on the star field.

758. The propulsive force resulting from the emission of gas and dust as the comet approaches the Sun (sometimes referred to as the nongravitational force and generally impossible to predict).

759. (d) Have orbits inclined at more-or-less random angles to the ecliptic plane.

760. ■ It arrived from outside the solar system or

   ■ Its orbit was converted from a near-parabolic or elliptical orbit to a hyperbolic one in a close flyby of Jupiter or another major planet (gravitational assist). See answer to problem 127.

761. (b) Have prograde orbits. (Comet Halley in a retrograde orbit is a noteworthy exception.)

762. 28 AU.

763. By a close flyby of Jupiter or another major planet (gravitational assist). See answer to problem 127.

764. As a comet approaches the Sun, its nucleus is progressively heated by sunlight and there is a correspondingly increased rate of emission of gas and dust.

765. (a) Solar heating of volatile constituents.

766. See answer to problem 767.

767. The value n = 2 for asteroids (inert bodies) corresponds to the inverse square dependence of the intensity of sunlight on R. In contrast, the effective reflecting area of comets increases as they approach the Sun because of the increasing emission of gas and dust from their nuclei; this effect causes n to be greater than 2.

768. 2 billion. If a sphere of radius R is disintegrated into n identical smaller spheres, each of radius r,

$$\frac{4\pi R^3}{3} = \frac{4\pi n r^3}{3} \quad \text{or} \quad n = \left(\frac{R}{r}\right)^3.$$

The ratio of the sum of the projected areas of the small spheres to the projected area of the large sphere is

$$\frac{n\pi r^2}{\pi R^2} = n\left(\frac{r}{R}\right)^2.$$

The increase in reflected sunlight (if no shadowing) is by this same factor, i.e., by the factor

$$\left(\frac{R}{r}\right)^3 \left(\frac{r}{R}\right)^2 = \frac{R}{r} = (n)^{1/3}.$$

769. All except methane, CH_4.

770. The two types of tails are ones consisting of fine particles of dust and ice (Type II) and ones consisting of ionized gas or plasma (Type I). The spectrum of light reflected by a dust tail is essentially that of sunlight. Dust tails extend outwards from the Sun and are bent backwards from (i.e., lag behind) the Sun-comet line. The spectrum of a gaseous tail is a composite of spectral emission lines characteristic of the composition of the ionized gas. Gaseous tails extend outward from the Sun along the Sun-comet line.

771. The magnetic field of the solar wind that flows radially outward from the Sun picks up the ionized cometary gas and carries it radially outward.

772. (a) Are essentially identical to the spectrum of sunlight.

773. (b) Provide important information about the composition of cometary material.

774. H, C, N, and O.

775. Characteristic spectra of H_2O, OH, H_2, and O_2.

776. (b) 16,000,000 miles.

777. About 32,000,000 miles.

778. (c) Ikeya-Seki.

779. A close encounter with a major planet.

780. (d) 3.81 AU.

781. (d) 75.1 years.

782. (b) 18 AU.

783. 35 AU.

784. (c) Lack of knowledge of its volatile constituents and physical structure.

785. 113 km s^{-2}.

786. (c) A meteor.

787. (c) The Earth's orbital motion is such as to increase the relative speed of the Earth and a meteoroid.

788. (c) Are small asteroids.

789. (b) Are probably caused by cometary debris.

790.

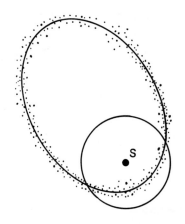

Meteor showers are attributed to finely divided cometary debris, distributed nonuniformly along the elliptical orbit of the parent body as sketched here. S represents the Sun and the circle, the orbit of the Earth. If the perihelion distance of the cometary orbit is less than 1.0 AU and if its plane lies in or near the ecliptic, two distinct meteor showers occur each year, centered on specific dates, as is evident in the diagram. In the more general case, even if the perihelion distance of the cometary orbit is less than 1.0 AU, one, two, or no showers per year occur depending on the eccentricity and the inclination to the ecliptic of the cometary orbit and on the orientation of its line of apsides. Quantitative ranges of these parameters can be specified for each case.

791. (a) Shower meteors.

792. The trails of meteors in this shower appear to be radiating from the constellation Perseus.

793. The trails of meteors in this shower appear to be radiating from the constellation Gemini.

794. Two cameras equipped with rotating or oscillating shutters, separated by a distance of the order of 50 km, and arranged so that their fields of view overlap at an altitude of 50 to 100 km.

795. 71 km.

796. (c) 630 miles radius.

797. It is impossible to make such an estimate from a single observing point.

798. Stone and iron/nickel.

799. Occasionally, a freshly fallen meteoric body is found at or near the point of projection of an observed meteor trail. This is an ideal case for reliable identification.

   However, most suspected meteorites are turned up in plowed fields, etc. and are chance finds. Validation of such samples depends upon (1) gross physical appearance, (2) chemical and mineralogical composition, (3) crystallization pattern (in iron/nickel samples), (4) presence of natural radioactive elements, and (5) radioactivity induced by cosmic ray bombardment.

800. (c) Fossilized skeletons of primitive animals. (False)

801. About 4 billion years.

802. 2.23×10^9 years.

803. 1.40/1.10/1.00.

804. The speed of a meteoroid with respect to the Earth has a lower limit of about 11 km s^{-1} because of the Earth's gravitational attraction. This lower limit corresponds to the capture of a meteoroid in an orbit similar to that of the Earth so that the relative speed before the encounter was essentially zero. At the other end of the range is a meteoroid in a retrograde elliptical orbit in the ecliptic plane with its perihelion at 1 AU. In this case the relative speed before encounter is $30 + 30\sqrt{2} = 72$ km s^{-1} and the speed of entering the Earth's atmosphere is $(72^2 + 11^2)^{1/2} = 73$ km s^{-1}. A typical observed value for shower meteors is 50 km s^{-1}.

805. (b) Result in the vaporization of much of its mass.

806. The Barringer crater near Flagstaff, Arizona. The diameter of this crater is 1,300 m and its depth is 180 m. A large quantity of iron fragments has been recovered but no single large object has been found.

807. During geologic history, craters on the Earth have been subjected to erosion and burial by glaciation, flowing water, dust storms, vegetative growth and decay, oceanic sedimentation, etc.—effects not present or far less important on the Moon, Mars, and Mercury. Also, a large fraction of the Earth's surface is covered by ocean. A meteor would make no crater if it fell in an ocean.

808. The area of the surface of the Earth is 5.1×10^{14} m^2. If one assigns a target area of 1 m^2 to each person, the total target area of 4×10^9 persons is 4×10^9 m^2. Thus about 10^{-5} of the Earth's surface is covered by people. It appears that the probability of some person being hit by a meteorite is of the order of 1/100 each year, i.e., one may expect that an average of one person will be struck by a meteorite in each 100 years (overhead shelter neglected).

☆ ☆ ☆

Radiations and Telescopes

809. (d) Gamma ray.

810. (c) Cosmic ray. (False)

811. Positively charged atomic nuclei (protons and the nuclei of helium and heavier elements) plus a minor admixture of electrons all having high energies, characteristically several billions of electron volts. Cosmic rays are present throughout the solar system and presumably throughout the universe. The processes by which they are accelerated to such high energies are thought to occur in supernova and in interstellar space.

812. (b) Roemer.

813. In its annual orbit around the Sun, the Earth alternately moves toward Jupiter and away from it. When the Earth is approaching Jupiter, the apparent orbital period of a Jovian satellite—as measured for example by the interval between successive occultations by the planet's disc—is less than its true sidereal period. The reason is that as the Earth-Jupiter distance decreases, the

light signal requires a lesser time to reach the Earth than if the Earth-Jupiter distance were constant. About six months later the situation is reversed and the apparent period is greater than the true period. The magnitude of this effect yields a measurement of the speed of light in terms of the speed of the Earth relative to Jupiter. It is, of course, desirable to cumulate the effect over many orbits in order to improve the accuracy of the determination.

814. (c) Finite speed of light.

815. During the period of revolution of Io (152,880 s) the Earth moves 4.59×10^6 km closer to Jupiter. An observer on the Earth notes that the interval between successive occurrences of a particular event in the orbit of Io (e.g., occultation by the planet) is $4.59 \times 10^6/300,000 = 15.3$ s less than expected. Thus, the apparent period of revolution of Io is less than its true value by this amount. The effect is of opposite algebraic sign about six months later (cf. answer to problem 813).

816. $300,000$ km s^{-1} ($= 186,000$ miles s^{-1}).

817. ■ The straightforward conceptual method for measuring the speed of light c is as follows.

Emit a brief pulse of light at the focal point of a lens or mirror so that the resulting parallel beam is directed to a plane mirror at distance D. Then by means of a photocell, detect the reflected pulse, measure the lapse of time Δt between emission and return of the pulse, and calculate the ratio $c = 2D/\Delta t$.

The experimental difficulty of making a precision determination of c by this method is suggested by the fact that with $D = 150$ meters, for example, Δt is about 1 microsecond. To obtain five significant figure precision, Δt must be measured to a precision of $\pm 1 \times 10^{-11}$ second.

■ The most successful classical method of measuring c, no-
tably by Michelson, employed a constant light source, a rotating multi-sided mirror as the effective time base, a distance
$D \approx 35$ km, and a measurement of the consequent deflection
of the focussed return beam that resulted from the rotation of
the mirror during the lapse of time Δt.

■ Modern indirect methods give the presently accepted value,
$c = 2.99792456(2) \times 10^5$ km s^{-1}.

818. 499 seconds or 8.32 minutes.

819. 63,200 AU per light year.

820. The Earth travels 2π AU per year. Therefore, the speed of light
is about 10,000 times the orbital speed of the Earth.

821. The intensity varies inversely as the square of the distance.

822. 1,600 times as great as at Pluto.

823. 900 times as great as at Neptune.

824. 27/1.

825. 27 meters diameter.

826. 55 meters diameter.

827. 1/16.

828. 10.4 times brighter than at the Earth.

829. The Pioneer 10 instruments are powered by four radioisotope terminal generators which derive electrical power from the decay products of plutonium 238. This electrical power is converted to heat. Also there are several smaller plutonium sources used for heaters at particular points in the spacecraft. Hence, the temperatures of all elements of the spacecraft are nearly independent of sunlight at its great distance from the Sun.

830. (a) The amount of sunlight that it absorbs.

831. About 2.7 degrees Kelvin.

832. 87 km s^{-1}.

833. (c) Assumes that the speed of light is independent of the relative motion of the source and the observer.

834. (d) The laws of physics are quite different in any coordinate system that is in motion with respect to the Earth. (False)

835. 0.976 c.

836. (d) 0.88 c.

837. (a) 4 meters.

838. 1.84 meters.

839. 330 meters.

840. 3 meters.

841. Identify one or more emission lines in the spectrum of the star with laboratory spectra of known elements. If λ is the wavelength of a chosen line in the stellar spectrum and λ_0, its wavelength in a laboratory spectrum, the radial speed v of the star relative to the Earth is given by the Doppler shift formula

$$\frac{v}{c} = \frac{\lambda - \lambda_0}{\lambda_0} .$$

If $\lambda > \lambda_0$, the star is moving away from the Earth (red-shift); if $\lambda < \lambda_0$, toward the Earth (blue-shift).

842. (d) Rotational period of Mercury. (See answer to problem 665.)

843. (d) 1.09 angstroms toward the violet.

844. 15 km s^{-1} away from the observer.

845. (c) The radial speed of a star.

846. In the Bohr theory only discrete, quite different orbits are permitted by the quantum principle. The same principle is applicable to planetary orbits but the differences between successive quantum states are minuscule and of no practical importance.

847. (d) A uniform gravitational field. (False)

848. (b) Wavelength.

849. (b) Diffraction.

850. (c) Refraction.

851. Refraction.

852. If a parallel beam of light is passed through a prism, the refraction (bending) of the beam is greater for short wavelengths and less for long wavelengths. Hence, the incident light is spread out into a spectral fan which reveals the distribution of intensity with wavelength.

853. (c) Refraction.

854. (d) RYGV.

855. (a) Refraction.

856. (c) Examining the spectrum of starlight.

857. (a) Measuring the Doppler effect.

858. (d) Coronagraph is used for producing artificial eclipses of the Sun.

859.

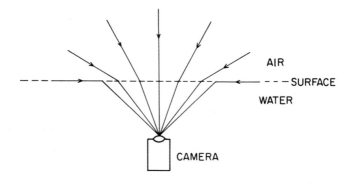

■ Rays of light from various objects in the upper hemisphere (e.g., half of the celestial sphere) are refracted at the air-water interface (n = 1.33) so as to reach the camera lens as shown in the diagram.

■ Note that the focussing properties of this wide-angle lens if immersed directly in water are determined by the indices of refraction of water and the material of the lens; hence they are substantially different than those for the same camera used in air. This effect can be avoided by using a watertight external window so that there is air on both external and internal sides of the lens.

860. (b) Real and inverted.

861.

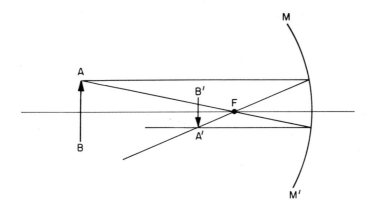

The image of point A is the intersection A′ of two representative rays. The first is parallel to the optical axis and after reflection from MM′ passes through the focus F. The second passes
through F and after reflection is parallel to the optical axis. The
image of point B is symmetrically located at B′ as shown. The
image A′B′ is real, inverted, and smaller than the object. The
two representative rays intersect at F, of course, but the full
bundle of rays from A converges only at A′.

862.

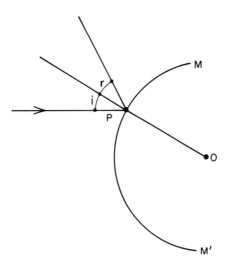

The plane of the diagram is the one defined by the incident ray
and point O.

■ Draw the radial line OP, which is normal to the surface of the mirror at P.

■ Measure the angle of incidence i between the incident ray and the line OP extended.

■ Draw the reflected ray on the opposite side of the line OP extended at a reflected angle r equal to i and in the plane defined by the incident ray and the normal.

863.

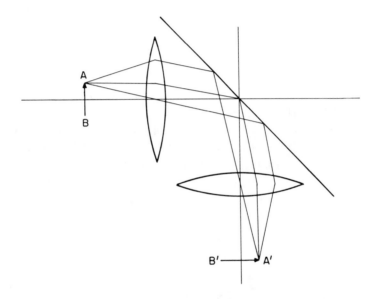

The image of A is A', the point of convergence of a bundle of rays emanating from A. Inasmuch as A is in the focal plane of the first lens, the bundle of rays is refracted by that lens so as to emerge as a parallel beam. It remains a parallel beam after reflection by the mirror and then strikes the second lens. The beam is focussed at A' in the focal plane of the second lens as shown by the convergence of three representative rays.

The image of B is B', found similarly.

864. (b) Virtual.

865.

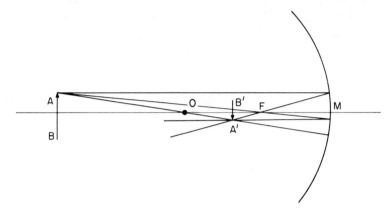

The approximate focal point F of a concave spherical mirror is located at half the radius of curvature from point M, i.e., FM = OM/2, where M is the intersection with the mirror of a line through the center of curvature O and the center of a distant object.

The image of A is A', the point of convergence of a bundle of rays emanating from A. Three representative rays are shown:

- One parallel to OM which is reflected through F;
- One through O which is reflected back along the same line; and
- One through F which is reflected parallel to OM.

The image of B is B', found similarly. The image A'B' is inverted, smaller than the object AB, and real.

866.

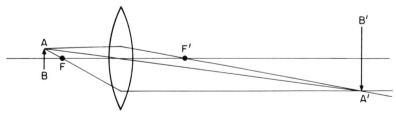

The image of A is A', the point of convergence of a bundle of rays emanating from A and passing through the converging lens. Three representative rays are shown:

■ One parallel to FF′, the optical axis, which is refracted so as to pass through F′;

■ One through the center of the lens which is undeviated; and

■ One through F which is refracted so as to be parallel to FF′.

The image of B is B′, found similarly. The image A′B′ is inverted, larger than the object AB, and real.

867. (c) 8.4 meters.

868. (c) 7 degrees.

869.   One minute of arc.

870. A golf ball.

871. (b) Increasing the diameter of its objective.

872. 113,000.

873. 18 million kilometers.

874. ■ A thin beam of light will be refracted into a circular path of radius r in a spherically symmetric planetary atmosphere having an index of refraction n(r) if a great circle of radius r + dr is traversed in the same time as one of radius r, viz.:

$$\frac{2\pi(r + dr)}{\left(\dfrac{c}{n + dn}\right)} = \frac{2\pi r}{\left(\dfrac{c}{n}\right)} , \qquad (1)$$

where c is the speed of light in a vacuum, or

$$\frac{dn}{dr} = -\frac{n}{r} . \qquad (2)$$

■ In a gas of density ρ

$$n = 1 + k\rho \qquad (3)$$

and

$$\frac{dn}{d\rho} = k , \qquad (4)$$

where k is a constant for a particular gas.

■ If the local altitude dependence of the density of the atmosphere of the planet is represented by a scale height H,

$$\frac{d\rho}{dr} = -\frac{\rho}{H} . \qquad (5)$$

■ Combining (2), (3), (4), and (5) one finds that

$$\frac{r}{H} = 1 + \frac{1}{k\rho} . \qquad (6)$$

For a planetary atmosphere $k\rho \ll 1.0$ and (6) becomes

$$r = \frac{H}{k\rho} . \qquad (7)$$

Equation (7) is the condition for the light path to be a great circle of radius r, where $r \gg H$.

■ This condition is not met anywhere in the Earth's atmosphere. However, in Venus' dense atmosphere, principally of CO_2, $k = 0.23$ cm^3 g^{-1}, and it is found that the condition for circularity is satisfied at an altitude of about 28 km above the planet's surface.

875. (a) The square of the diameter of its objective.

876. The diameter of the objective (lens or mirror).

877. 1.7 inches.

878. (d) 10,000 times greater than that of the human eye.

879. (d) 0.7 mm.

880. (b) 2 cm.

881. (b) Y is faster than X.

882.

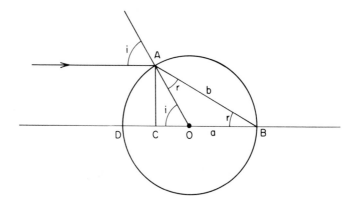

$$CD = a(1 - \cos i)$$
$$CB = b \cos r = 2a \sin \left(\frac{180 - i}{2} \right) \cos r$$
$$CD + CB = 2a$$
$$n = \frac{\sin i}{\sin r}$$

■ As evident from the diagram, the value of n such that the rays of a given i intersect at the surface of the sphere is obtained from the geometric relation

$$\cos r = \frac{1 + \cos i}{2 \sin \left(\frac{180 - i}{2} \right)} \tag{1}$$

and the refraction equation

$$n = \frac{\sin i}{\sin r}, \tag{2}$$

where r is the angle of refraction as the rays enter the sphere.

■ Sample results are as follows:

i	n
10 degrees	1.992
20	1.970
30	1.932
40	1.879
50	1.813
60	1.732
70	1.638
80	1.532
90	1.414

883. ■ A ray of light emerging from glass is refracted away from the normal to the glass-air interface at an angle r such that

$$\frac{\sin r}{\sin i} = n , \tag{1}$$

where i is the angle of incidence and n, the index of refraction of the glass. As r approaches 90 degrees, the emerging ray skims the interface. This case corresponds to

$$\sin i = \frac{1}{n} .$$

■ A typical value of n is 1.5 and the corresponding value of i is 42 degrees. For lesser values of i, the ray escapes into the air. For greater values of i, the ray is reflected as though the interface were a mirror. This scheme of internal reflection has many uses in astronomical instruments and in corner reflectors (cf. problem 894).

884. (b) 33 inches.

885. 0.63 inch.

886. 115 cm.

887. The image with telescope B is four times as bright as the image with A.

888. (e) 1 mile.

889. (a) $1/100^{\text{th}}$ of a second.

890. Y produces an image that has eight times the linear dimension of the one produced by X whereas X produces an image that is four times as bright as is the one produced by Y.

891. (a) 144.

892. $6/100^{\text{th}}$ of a second.

893. (c) The light-gathering power of the telescope is reduced by about 6 percent.

894. The proposition is obviously true for the special cases in which a total of only one or two reflections occur.

■ In the general case, three reflections occur, one with each of the three mirrors. Let the mirrors be the XY, XZ, and YZ planes, respectively, and let the entering ray be specified by the unit vector

$$\vec{A}_0 = a_x i + a_y j + a_z k \ ,$$

where a_x, a_y, and a_z are constants such that $a_x{}^2 + a_y{}^2 + a_z{}^2 = 1$; and i, j, and k are unit vectors, parallel to the X, Y, and Z axes, respectively.

■ After reflection by the XZ plane

$$\vec{A}_1 = a_x i - a_y j + a_z k \ .$$

■ After subsequent reflection by the YZ plane

$$\vec{A}_2 = -a_x i - a_y j + a_z k \ .$$

■ And after subsequent reflection by the XY plane

$$\vec{A}_3 = -a_x i - a_y j - a_z k \ .$$

Hence,

$$\vec{A}_3 = -\vec{A}_0 \qquad \text{Q.E.D.}$$

The order in which the three reflections occur is immaterial.

■ Two arrays of such corner reflectors (using internal reflections within a prism) were left on the surface of the Moon by Apollo astronauts. These arrays have been illuminated by a pulsed laser beam from a terrestrial telescope. The observed round-trip flight time of a laser pulse measures the precise distance (within a few cm) from the telescope to the reflector. Other such arrays have been flown on satellites for precision geodetic studies. Common uses of corner reflectors include radar reflectors (of wire mesh) on weather balloons and navigational buoys; highway and auto-mobile optical reflectors; and targets for surveyors.

895. (a) 64 times as great as that of a 25-inch telescope.

896. (a) The A/B exposure time ratio is 6.25.
    (b) The diameters of the images are A, 0.52 inch; B, 4.2 inches.

897. (a) Erect and larger than the image formed by the objective.

898. (d) Largest available focal length.

■ The surface brightness of the image of the general luminous glow is proportional to the inverse square of the f/no, i.e., proportional to $(d/F)^2$, where d is the diameter of the objective and F, its focal length. The amount of light that images a star is proportional to d^2; the dimensions of the image of a star, caused by atmospheric "seeing" are proportional to F; hence the area of the star's image is proportional to F^2. On these considerations there appears to be no advantage to any one of the suggested choices. This conclusion is in fact true for an array of linear detectors of infinitesimal dimensions.

■ However, for a photographic plate, the diameter of a given element is of the order of 20 microns and such a grain is a nonlinear detector, i.e., after some specified exposure it is completely black and gets no blacker with further exposure. Hence, it is desirable to spread the image over more than one grain in order to detect faint stars. The result is that it is advantageous to use a telescope having the largest available focal length to reveal faint stars in the presence of sky glow.

899. (a) The prime focus of a telescope is never accessible to observing equipment. (False)

900. (a) An astronomical object can be tracked by rotating the mount about only one axis, the polar axis (ideally).

901. (d) 24^h 50^m.

902. 6 times that with B.

903. 1.9 km.

904. (d) 2 ft.

905.

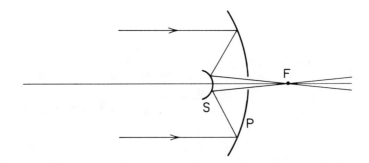

The essential optical elements of a Cassegrain telescope are a concave, usually spherical, primary mirror P and a convex secondary mirror S located closer to P than the latter's focal point. Two parallel rays from a remote source are sketched. After reflection from the two mirrors they converge at the focal point F of the composite optical system.

906. (b) Atmospheric "shimmer".

907. (c) Reduce the brightness of stellar images by four percent.

908. (a) The larger the objective, the greater is the resolving power;
     (b) The greater the focal length, the larger is the primary image; and
     (e) The larger the objective, the brighter is the primary image.

909. There are four advantages, the first being the most noteworthy:
     - The full spectrum is accessible, free of atmospheric absorption—an advantage especially important in the ultraviolet.
     - Atmospheric "shimmer" ("seeing") is eliminated and better resolution is possible.
     - The background due to natural airglow, moonlight, and artificial light scattered in the atmosphere is eliminated.
     - There are no clouds between the telescope and the objects under study.

910. (c) The accessibility of the ultraviolet portion of the spectrum.

911. (b) A hot solid (or hot, dense gas).

912. (a) A hot, tenuous gas.

913. (d) Hotter than the Sun.

914. Sunlight passes inward through the atmospheric gas to the surface of the planet, is partially reflected at the surface and passes back outward through the atmospheric gas. Absorption lines and bands in the resulting spectrum are compared to those characteristic of various molecules in order to identify them in the planet's atmosphere.

915. (c) Their characteristic absorption lines in the solar spectrum.

916. 3,600 degrees Kelvin (Wien's Law).

917. (b) Its atmosphere consisted principally of hot, tenuous gas.

918. In general, a radio telescope is an antenna having a directional dependence of its sensitivity to radio signals and a system of detection and recording equipment. In astronomy, the principal element of a typical radio telescope is a concave paraboloidal reflector having a diameter of the order of many meters.

919. A beam of radio waves from a distant source, if parallel to the axis of the paraboloid, is reflected exactly through the focal point. This is not true for a spherical concave reflector.

920. (c) 4.6 times as much radio power as does the smaller.

921. (d) Six degrees of arc.

922. 0.12 degree of arc.

923. (d) An optical diffraction grating.

924. Determination of
    - Absolute value of the astronomical unit, in terms of the speed of light;
    - Rotation period of Mercury;
    - Rotation period of Venus;
    - Temperatures of the surfaces of the Moon, Mars, Mercury, and Venus;
    - Synchrotron emission by relativistic electrons in Jupiter's magnetosphere;
    - Detailed topography of the surfaces of the Moon, Mars, and Venus;
    - Nonthermal emissions of Jupiter, Saturn, Uranus, and Neptune; and
    - Plasma phenomena in planetary magnetospheres and the interplanetary medium.

    Also
    - Precise tracking of spacecraft.
    - Radio telemetry from spacecraft.
    - Commands to spacecraft.

Recommended References

Abell, G. O., D. Morrison, and S. C. Wolff, *Exploration of the Universe*, 6th Ed., Saunders College Publishing, Philadelphia, 1993.

Allen, C. W., *Astrophysical Quantities*, 3rd Ed., The Athlone Press, London, 1973.

The Astronomical Almanac, an annual publication of the U.S. Naval Observatory and the Royal Greenwich Observatory, U.S. Government Printing Office, Washington, D.C.

SC 1 Constellation Chart—Equatorial Region—Epoch 1925, Sky Publishing Corporation, Cambridge, MA.

SC 2 Constellation Chart—North Circumpolar Region—Epoch 1925, Sky Publishing Corporation, Cambridge, MA.

Sky and Telescope, a monthly publication of the Sky Publishing Corporation, Cambridge, MA.

Norton's 2000.0 Star Atlas and Reference Handbook, 18th Ed., J. Ridpath, Editor, Longman Scientific and Technical, Essex, England, 1989.

Observer's Handbook, R. L. Bishop, Editor, an annual publication of The Royal Astronomical Society of Canada, Toronto.